EXPERIMENTAL
ELASTICITY

EXPERIMENTAL ELASTICITY

A MANUAL FOR THE LABORATORY

BY

G. F. C. SEARLE, Sc.D., F.R.S.

UNIVERSITY LECTURER IN EXPERIMENTAL PHYSICS
AND
DEMONSTRATOR IN EXPERIMENTAL PHYSICS
AT THE CAVENDISH LABORATORY

SECOND EDITION

CAMBRIDGE
at the University Press
1933

CAMBRIDGE UNIVERSITY PRESS
Cambridge, New York, Melbourne, Madrid, Cape Town,
Singapore, São Paulo, Delhi, Mexico City

Cambridge University Press
The Edinburgh Building, Cambridge CB2 8RU, UK

Published in the United States of America by Cambridge University Press, New York

www.cambridge.org
Information on this title: www.cambridge.org/9781107664227

© Cambridge University Press 1933

First Edition 1908
Reprinted 1920
Second Edition 1933
First published 1933
First paperback edition 2013

A catalogue record for this publication is available from the British Library

ISBN 978-1-107-66422-7 Paperback

PREFACE

THE present volume has its origin in the manuscript notes which I have prepared from time to time for the use of the students attending my class in practical physics at the Cavendish Laboratory. When, in 1890, I was appointed to my present post of Demonstrator in Experimental Physics, I found that the then existing text-books of practical physics did not entirely meet the needs of the students, partly because they did not, as a rule, show how the formulae required in the experimental work are derived from the principles of the subject. The students themselves added to the difficulty, for their ideas as to those principles were often indistinct. I was thus led to devise some experiments intended to illustrate principles as simply and directly as possible. I also wrote notes explaining how the necessary formulae are obtained from the principles involved in those experiments and describing in detail how the practical work is to be conducted. The students showed very kind appreciation of these earlier notes and thus I was encouraged to prepare others; this work has proved so interesting that I have continued it, as opportunities have occurred, with the result that at the present time the students attending my practical class rely mainly upon these manuscript notes for the necessary instructions.

Many of the students have made almost complete copies of some hundreds of pages of manuscript and have perhaps learned more in that way than by merely reading a printed book containing the same matter. But the plan of using manuscript notes has numerous disadvantages. For instance, the limited number of copies of any one manuscript makes it difficult to arrange for more than two or three students to do the same experiment at one time and often prevents the students from preparing themselves beforehand for the experiments assigned to them. There is, besides, the risk of the loss or the destruction of the manuscripts themselves. For the safety of the manuscripts I have relied on the consideration of the students and this has hardly ever failed.

To throw together into a small volume the manuscripts dealing with one branch of physics would seem an easy task. But the result would hardly be satisfactory, for some of the earlier manuscripts require revision in the light of later experience, while many of the manuscripts contain mathematical arguments which are repeated in others of the series. This repetition was necessary for the practical working of the class but would be intolerable in a book.

For these and other reasons I decided that it would be more satisfactory to rewrite the whole of the manuscripts, and to arrange the material, with additions, in the form of a series of small textbooks, in which a fairly full account of the mathematical treatment should accompany a detailed description of the experimental work.

To make a beginning, the present volume is published and this, I hope, will be followed in a few months by a similar volume on Experimental Optics. I hope, if life and health be given me, to complete the scheme by writing volumes on Mechanics, on Electricity and Magnetism and on Heat and Sound.

The present volume cannot lay claim to any sort of completeness. Its purpose is simply to give the substance of my course of instruction in the subject in a form which may be useful to students at the Cavendish Laboratory and elsewhere.

The first chapter contains an account of the elements of the mathematical theory of elasticity, with one or two necessary propositions in thermodynamics. In the second chapter will be found the mathematical solutions of some problems which make their appearance in several experiments. The uniform bending of rods and blades is discussed rather fully, but I was anxious to make the arguments apply to small *finite* curvatures as distinguished from merely infinitesimal curvatures. To the preparation and revision of this chapter, Dr L. N. G. Filon has contributed so much from his store of expert knowledge of the mathematical theory of elasticity that the chapter is almost more his work than mine.

The third chapter contains descriptions of a number of experiments together with such necessary mathematical discussions as are not given in the first two chapters. Each description is followed by a practical example giving detailed arithmetical results taken from an actual experiment; these examples may perhaps assist students in recording their own observations.

Some notes bring the book to a close. The last of these contains hints on practical work in physics; the rest deal mainly with a few dynamical theorems which experience suggests may be useful to students who have not received a mathematical training.

Most of the apparatus required for the experiments is of a simple description. Though in some cases accuracy would be gained if the apparatus had less of the " home made " character and more of the engineer's workmanship, this roughness of the appliances is not a serious disadvantage to the students who use the apparatus at the Cavendish Laboratory. Those who afterwards make physics a part of their work, either as teachers or as investigators, will probably have to struggle on with a good deal of "home made" apparatus. To the rest, who distribute themselves over very wide fields of human activity, a knowledge of principles is of greater value than an acquaintance with the details of highly finished instruments.

In the design of the apparatus I have often been aided by Mr W. G. Pye and by Mr F. Lincoln, the past and present instrument makers at the Cavendish Laboratory, and by their assistants.

To assist those teachers who may not be able to construct the apparatus for themselves, I have authorised Messrs W. G. Pye and Co., of Cambridge, to supply apparatus made to my designs. I have done this because, in some cases, instrument makers, without consulting me, have connected my name with apparatus in which they have made "improvements" of doubtful value.

I owe much to the many generations of students who have attended my class. Their never failing enthusiasm has been a source of much encouragement to me, and the honest work and the satisfactory progress of the great majority has been a real reward.

I also owe much to the kindness of those who have assisted me as demonstrators during eighteen years, and especially to the unwearying help which my oldest colleague, Mr T. G. Bedford, has given in many ways for many years.

This volume owes much to the generous help rendered me by friends. The proofs have been read and criticised by Mr Bedford; his knowledge of physical principles, of the work of teaching the experimental methods described in this book and of the difficulties of students makes his aid of great value. Dr Alexander Russell, who has had a long experience of students' work, has made many helpful criticisms upon the proofs. Dr L. N. G. Filon has spent much labour upon the first and second chapters, and Mr W. C. D. Whetham, F.R.S., a former colleague, has given editorial assistance.

Mr D. C. Jones of Pembroke College, Mr P. D. Innes of Trinity College and my wife have assisted in preparing the manuscript for the press, while Mr A. J. Bamford of Emmanuel College has helped in the revision of the proofs. To all these, as well as to those who have aided in minor ways, my thanks are given.

The following words, from Psalm cxi (*v.* 2), which are carved on the gates of the Cavendish Laboratory, shall end this preface: *Magna opera Domini: exquisita in omnes voluntates ejus.*

G. F. C. S.

August, 1908.

PREFACE TO THE SECOND EDITION

IN this edition two Notes have been added. Note XI deals with a detail of EXPERIMENT 6. Note XII is devoted to the theory of the *infinitesimal* uniform bending of a rod, and it is shown that $G\rho$ tends to the *limit EI* as the curvature $1/\rho$ tends to zero. The treatment of the uniform bending of rods and blades given in Chapter II in §§ 27 to 38 is more thorough than that given in Note XII, for it shows what conditions must be satisfied if $G\rho$ is to be a good approximation to EI when the bending is *finite* but still small, and it brings out the important distinction between the finite bending of a rod and the finite bending of a blade. This distinction is, in the nature of things, outside the scope of Note XII. The theory of Chapter II is necessarily lengthy and may be found difficult by students of small experience and perhaps tedious by those who are keen to "cover the ground" as fast as possible. It is hoped that Note XII will help the beginners and encourage them to read Chapter II.

A few misprints have been corrected and a few sentences have been modified.

G. F. C. S.

CAVENDISH LABORATORY,
CAMBRIDGE.
20 *September*, 1933.

CONTENTS

CHAPTER I

ELEMENTARY THEORY OF ELASTICITY

CHAPTER II

SOLUTIONS OF SOME SIMPLE ELASTIC PROBLEMS

CHAPTER III

EXPERIMENTAL WORK IN ELASTICITY

CHAPTER I.

ELEMENTARY THEORY OF ELASTICITY.

1. Introduction. The application of a system of forces to a solid body causes a deformation corresponding to the character of the system of forces; for example, a pull causes an extension while a couple causes a twist in a wire. But the simplest observations on the stretching or bending of a piece of copper wire are sufficient to show that, even though the forces are not so great as to break the body, they may still be great enough to produce changes of form, which do not entirely pass away when the forces are removed. The effects of forces of this character are of great importance in many industries. The moulding of clay in pottery work, and the forging, stamping, wire drawing and cutting of metals are familiar instances of such effects.

When the forces are less intense, the body may so nearly recover its original form, on the removal of the forces, that careful observations are required to show that the recovery is imperfect.

It is, therefore, natural to assume that, if the forces be small enough, the body will completely recover its original form on their removal. This is equivalent to saying that the form of a body depends only on the forces which act on it at the time, and not upon those which have ceased to act. The assumption that the forces have no after-effects is of great importance, because it renders the mathematical treatment of the subject comparatively simple. The assumption is probably not strictly true for any substance, but for many substances it is so near the truth that, for practical purposes, it may be regarded as exactly true.

2. Hooke's law. Though Robert Hooke was the first to publish a definite statement as to the relation between small forces and the changes of form due to them, yet it is probable that most of the persons who had made any practical use of springs had at least a working knowledge of that relation.

In 1676 Hooke published the statement:

" The true Theory of Elasticity or Springiness, and a particular Explication thereof in several Subjects in which it is to be found: And the way of computing the velocity of Bodies moved by them. ceiiinossssttuu."

In 1678 he gave the key to this anagram in the words:

"About two years since I printed this Theory in an Anagram at the end of my Book of the Descriptions of Helioscopes, viz. ceiiinossssttuu, id est, Ut tensio sic vis; That is, The Power of any Spring is in the same proportion with the tension thereof: That is, if one power stretch or bend it one space, two will bend it two, and three will bend it three, and so forward."

The proportionality between the applied forces and their effects is known as Hooke's law and forms the basis of the mathematical theory of the subject. In this theory it is further assumed that when two or more sets of small forces act on a body, each set produces the same effect as if the other set or sets were not acting. This assumption is, however, only a natural extension of Hooke's law.

Within the range where Hooke's law holds, we may speak of the body as being *perfectly elastic*.

If the forces acting on the body be increased, a more or less definite point is reached where Hooke's law begins to fail. When Hooke's law fails, we may say that the *elastic limit* of the body has been passed.

3. Necessity for a theory of elasticity. In §§ 1 and 2 we have taken an elastic body as a whole and have not considered the actions between its parts. The results which can be obtained in this way are sufficient for some purposes. Thus, if we make a helical spring of steel wire, we can use it as a spring balance and can, by experiments with known masses, graduate a scale so that the balance shall indicate the mass of any body suspended from it, and this can be done without any reference to the complex actions which occur within the steel itself.

The process here indicated may be considerably extended, for, if we take a series of bodies of similar form and of the same material, and subject them to similar sets of forces, we can, from these experiments, deduce laws which would enable us to *predict* the behaviour of another body, if of similar form and of the same material, when subjected to a similar set of forces. Thus, we should find by experiment that, when a wire of length l and cross section A is subjected to a pull F, the increase of length λ is given by

$$\lambda = \frac{plF}{A},$$

where p is a constant depending upon the material. From this equation the increase of length produced in any given wire of that material by any given pull could be calculated. Similarly, we could find by experiment that the total twist θ, produced by a couple G in a circular rod of radius r and length l, is given by

$$\theta = \frac{qGl}{r^4},$$

where q is a constant depending on the material.

Though the results obtained in this way would be of great practical utility, they would fail to provide a means of *calculating* the effect of any given set of forces on any given body. Thus, experiments on the torsion of a rod of circular section would give no information as to the twist which a given couple would produce in a rod of the same material but of rectangular section.

It thus becomes evident that we need a theory of elasticity, by which we can calculate mathematically, if we have sufficient skill, the effect of any given set of forces on any given body, when we have found the "elastic constants" of the material by experiments made upon specimens of the material. We shall, therefore, devote this chapter to the elements of such a theory.

The material will be supposed to be isotropic, i.e. to have the same properties in all directions, and to be homogeneous, i.e. to have the same properties at all points.

4. Action and reaction between two parts of a body. Let the body be divided into two parts A and B by a mathematical surface. Each part will in general exert a set of forces on the other, and the whole action of A on B and of B on A is due to forces acting between the molecules of A and those of B. These forces may be divided into two classes. In the first class are those that are sensible at more than molecular distances; this class includes gravitational, electric, and magnetic actions. In the second class are those forces which are sensible only within molecular distances. We shall speak of these last forces as due to molecular actions.

Now, it is only those molecules which lie on one side of the dividing surface within a distance of about 10^{-8} cm. from the surface, which have any appreciable effect, by molecular action, on those on the other side. But, in the layer corresponding to one square centimetre of the surface, there are, in the case of a solid or a liquid, about 10^{16} molecules and thus an element of the layer of only one millionth of a square millimetre in area contains about 10^8 molecules. It is evident, therefore, that, if the elements of area, which we consider, are not very small compared with a millionth of a square millimetre, the multitude of small forces arising from molecular actions may be considered as blending together into a force continuously distributed over the element of area. In other words, the forces which act on the part of the body on either side of the surface, and are due to molecular action, may for all practical purposes be replaced by a force continuously distributed over the surface, the forces on the two parts being everywhere equal and opposite. What we have done here is equivalent to replacing the molecularly-built body by one of absolutely continuous structure.

In general, the part A is acted on not only by the molecular actions due to B, which are included in the second class, but also by those forces due to B which are included in the first class. In addition, the part A experiences forces due to the action of other bodies, as when it is pulled by a string or is attracted to the earth through gravitation.

If we apply Newton's laws of motion, we find that the rate of increase of the momentum of A in any direction is equal to the resultant in the same direction of all the forces acting on A and that the rate of increase of the angular momentum of A about any fixed axis is equal to the moment about the same axis of all the forces acting on A *.

In many cases the part A is at rest and then it follows (1) that the resultant force arising from the molecular actions of B is in equilibrium with the resultant of the remaining forces which act on A, and (2) that the moment about any axis of the molecular actions of B is in equilibrium with the moment about the same axis of the remaining forces which act on A. These results are frequently used in experimental work.

5. Stress. The word *stress* is often used in a general sense in connexion with the action of forces, but in this book it will be used only in a definite mathematical sense. Thus: If any elementary area be drawn in the body, the parts of the body on either side of the area exert equal and opposite forces on each other by molecular actions arising from molecules in the immediate neighbourhood of the area. The ratio of either of these forces to the area is called the *stress*. The stress may be normal or tangential to the area, or may be inclined at any angle to a line normal to the area.

When the stress is normal to the area, it is called a pressure or a tension according to its direction, and when it is tangential to the area, it is called a shearing stress.

It is shown in § 7 that, in the case of a hydrostatic pressure, where for *every* elementary area containing a given point the stress is normal to the area, the *magnitude* of the stress at that point is independent of the direction of the normal to the area, but, in the general case, the magnitude of the stress and its inclination to the normal will both depend upon the direction of that normal.

6. Measurement of stresses. The numerical value of a given stress depends upon the units of force and of area which we employ. To avoid errors, the student should be careful to state correctly the unit of force employed and to specify the unit adopted for the measurement of areas. In the C.G.S. system, which is used in this book, the stresses are measured in terms of a unit stress of one dyne per square centimetre.

* See Note II.

As a simple example, suppose that a vertical wire 1·32 milli-metres in diameter supports a mass of 3·5 kilogrammes in a locality where $g = 981$ cm. sec.$^{-2}$, and that the stress is required for a plane cutting the axis of the wire at right angles. The total force acting across the plane is $3\cdot5 \times 1000 \times 981$ or $3\cdot434 \times 10^6$ dynes, while the area of section is $\pi (0\cdot066)^2$ or $1\cdot368 \times 10^{-2}$ square cm. Hence the stress (assumed uniform) is a normal one and its magnitude is

$$\frac{3\cdot434 \times 10^6}{1\cdot368 \times 10^{-2}} = 2\cdot510 \times 10^8 \text{ dynes per square cm.}$$

If the normal to the plane be inclined at an angle θ to the axis of the wire, the area of section is $1\cdot368 \times 10^{-2} \times \sec \theta$. But the total force is still $3\cdot434 \times 10^6$ dynes in a vertical direction. Hence, in this case, the stress makes an angle θ with the normal to the plane and its magnitude is $2\cdot510 \times 10^8 \times \cos \theta$ dynes per square cm.

7. Hydrostatic pressure. When, for every elementary area containing a given point, the stress is normal to the area, there is said to be a hydrostatic pressure at the point. Now consider the matter contained in an elementary tetra-hedron $OABC$ (Fig. 1). Let the edges OA, OB, OC be mutually perpendicular and let X, Y, Z be the stresses on the faces OBC, OCA, OAB. Let S be the area of ABC and α, β, γ be the angles between this plane and the planes OBC, OCA, OAB. The force required to give the enclosed matter any acceleration it may have is proportional to the *cube* of the linear di-mensions of the tetrahedron, as is also any force arising from gravity. But the forces

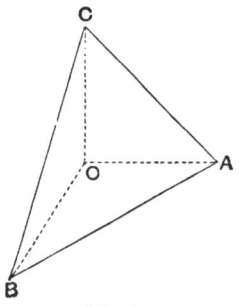

Fig. 1.

due to the stresses on the bounding planes are proportional to the *square* of the linear dimensions and hence, by taking the tetra-hedron small enough, the forces due to the stresses may be made as great as we please compared with the other forces. Thus, in the limit, we need only consider the stresses.

Let P be the stress on ABC. Then, since the only forces which have components parallel to OA are a force $P \cdot S$, acting normally to ABC, and a force $X.OBC$ or $X.S$ cos α, acting normally to OBC, we have

$$PS \cos \alpha = XS \cos \alpha,$$

with two similar equations.   Hence

$$P = X = Y = Z,$$

so that the stress is independent of the direction of the normal to the area.

8. Strain. Suppose that, before the forces are applied to an elastic body, three series of planes are drawn in the body so as to divide it into infinitesimal cubes.   When the forces are applied, the portions of matter originally in these cubes will, in general, be changed in shape so that they are no longer of cubical form; they will also be changed in volume.   The change of form, including the change of volume, which occurs in any elementary cube may be called, in general terms, the *strain* at that part of the body.   But it is evident, from this description, that, in the general case, no single quantity is sufficient to measure the change of form which occurs in any elementary cube, and thus we see that, in general, more than one quantity is required to specify the strain. For the mathematical treatment of strains in general the reader is referred to treatises on the mathematical theory of elasticity.

For our purpose it will be sufficient to consider two fundamental strains and some simple strains which can be built up from them.

9. Expansion and compression. When the strain is such that any elementary cubical portion of the body remains cubical, although changed in volume, the strain is called an expansion or a compression, according as the volume of the cube is increased or diminished by the strain.   In either case the strain is measured by the change of volume per unit volume.

Thus, if the application of pressure to the surface of a piece of steel reduce the volume of each cubic cm. by 2×10^{-12} cubic cm. without otherwise changing its form, the strain is a uniform compression amounting to 2×10^{-12} c.c. per c.c.   It will be noticed that the numerical value of the compression is independent of the unit of volume employed.   But care must be taken to measure both the original volume and the diminution of volume in terms of the *same* unit.

10. Shear. In the case of compression there is a change of volume without any further change of shape. We now go on to examine the simplest case of change of shape without change of volume.

Consider two parallel planes A, B drawn in a body at the distance h apart and suppose that all the particles in the plane A remain fixed in position. If, now, every particle in the plane B be moved *in that plane* through the same distance and in the same direction, the part of the body between the planes is said to be *sheared*. If the displacement of any particle between A and B be parallel to that of the particles in B and proportional to the distance of the particle from the plane A, the strain is called a uniform shear. If we take a rectangular block having one face in each of the planes A and B, this block will be strained into a parallelepiped of equal volume, since both the area of the base and the height of the block remain unchanged. Thus a uniform shear does not change the volume of the body.

A plane through any point P between the planes A and B and parallel to them is called the plane of the shear at P.

To measure the magnitude of the shear, we take a cubical block having one face in each of the planes A and B and four edges parallel to the direction of the displacement of the particles in B. Thus the faces in the planes A, B remain squares, the faces normal to the direction of displacement are strained into rectangles, and the remaining faces are distorted into parallelograms. If the angles of a distorted face $A_1A_2C_2C_1$ (Fig. 2) are no longer all equal to $\frac{1}{2}\pi$ but are $\frac{1}{2}\pi + \theta$ and $\frac{1}{2}\pi - \theta$ radians, as indicated in the figure, the strain is said to be a shear of θ radians.

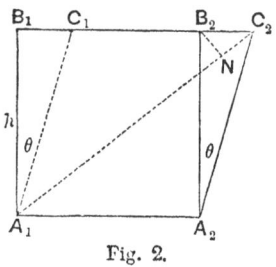

Fig. 2.

11. Maximum shear in actual experiments. For all metals Hooke's law only holds for small shears and ceases to apply when the shear exceeds $\frac{1}{300}$ radian or one third of a degree, either because the metal breaks before this shear is reached (e.g. hard steel) or because it flows (e.g. lead). In ordinary experiments for finding the elastic constants the shear need never exceed $\frac{1}{1000}$ radian.

12. Results for infinitesimal shears. We shall now obtain some useful results, which are approximately true for small shears and accurately true for infinitesimal shears. From Fig. 2

$$B_1C_1 = B_1A_1 \tan \theta = h \tan \theta.$$

But, when θ is less than $\frac{1}{1000}$ radian, we may put $\tan \theta = \theta$ with an error of less than one in three millions and thus may write $B_1C_1 = h\theta$.

Since $\quad A_1C_1 = (A_1B_1{}^2 + B_1C_1{}^2)^{\frac{1}{2}} = h(1 + \tan^2 \theta)^{\frac{1}{2}}$,

we find, on expanding by the binomial theorem and replacing $\tan \theta$ by θ, that

$$A_1C_1 = h(1 + \tfrac{1}{2}\theta^2 - \ldots\ldots).$$

Hence the shear does not alter the length of the edges A_1B_1, A_2B_2 by more than $\frac{1}{2}\theta^2$ cm. per cm. If $\theta = \frac{1}{1000}$, the change does not amount to one part in 2,000,000.

In technical mathematical language we may say that θ, the shear, is a small quantity of the first order and that $\frac{1}{2}\theta^2$, the elongation, i.e. the increase of length per unit length, of the edges A_1B_1, A_2B_2, is a small quantity of the second order.

In the mathematical theory the strains are supposed to be infinitesimal. In this case the difference between A_1C_1 and A_1B_1 is to be neglected, and then we may say that the edges of the cube are unchanged by the strain. We have already seen in § 10 that the volume of the cube is unchanged. The two statements are inconsistent when the shear is finite, but they become consistent when the shear becomes infinitesimal.

If B_2N be drawn perpendicular to A_1C_2, we may take NC_2 as the increase in length of the diagonal A_1B_2. Now the angle NC_2B_2 is ultimately equal to $\pi/4$ and thus, in this case,

$$NC_2 = B_2C_2 \cos \pi/4 = h\theta/\sqrt{2} = \tfrac{1}{2}\theta \cdot h\sqrt{2}.$$

Similarly, the length of the diagonal A_2B_1 is diminished by $\frac{1}{2}\theta \cdot h\sqrt{2}$. But $h\sqrt{2}$ is the length of the original diagonals, and thus we see that the shear θ has lengthened one diagonal and shortened the other by $\frac{1}{2}\theta$ cm. per cm.

The lengths A_1C_1, A_1A_2 only differ by a small quantity of the second order, and hence the diagonals A_1C_2 and A_2C_1 may be considered as intersecting at right angles. The strain has therefore not changed the angle between the diagonals, though it has turned each diagonal through an angle equal to B_2N/A_1B_2 in the same direction. Since $B_2N = NC_2 = \frac{1}{2}\theta \cdot h\sqrt{2}$ and since $A_1B_2 = h\sqrt{2}$, it follows that this angle is $\frac{1}{2}\theta$.

Since the strain is uniform, all lines in the block parallel to the diagonal A_1B_2 will be lengthened by $\frac{1}{2}\theta$ cm. per cm., and similarly all the lines parallel to A_2B_1 will be shortened by the same amount. Each set of lines will intersect the other at right angles after as well as before the straining, though each set will be turned through $\frac{1}{2}\theta$ in the same direction.

Hence, a uniform shear of θ radians is equivalent to a uniform contraction of $\frac{1}{2}\theta$ cm. per cm. in a direction inclined at 45° to the plane of the shear (§ 10), superposed on a uniform extension of $\frac{1}{2}\theta$ cm. per cm. at right angles to the contraction.

13. Bulk modulus or volume elasticity. Suppose that, at every point within a body of homogeneous and isotropic matter, the stress is a uniform hydrostatic pressure of p dynes per square cm. This will evidently compress each elementary cube of the body in the same proportion, and hence the strain will be a uniform compression. If we take the triangle ABC (Fig. 1) to be an element of the surface of the body, we see that, to secure the equilibrium of the elementary tetrahedron $OABC$, a uniform pressure p must be applied to the surface. Conversely, we may conclude that a uniform pressure p applied to the surface of a body of homogeneous and isotropic matter gives rise to a hydrostatic pressure p throughout the body and produces a uniform compression.

This result does not apply when the body contains a cavity, unless a pressure p be applied to the walls of the cavity as well as to the outer surface of the body.

It is found by experiment that, so long as the pressure is not too great, the compression is proportional to the pressure, and thus the ratio of the pressure to the compression may be regarded as an " elastic constant " of the material.

The ratio of the pressure to the compression is called the *bulk modulus* or *volume elasticity* of the substance and is denoted by k. Since the compression is a pure number, the bulk modulus is measured in the same units as the pressure, i.e. in dynes per square cm.

If a pressure p cause the volume of the body to diminish from v to $v - w$, the compression is w/v, and hence the bulk modulus is given by

$$k = \frac{\text{stress}}{\text{strain}} = \frac{\text{pressure}}{\text{compresssion}} = \frac{pv}{w} . \qquad \ldots\ldots\ldots\ldots(1)$$

As an example of the use of this formula, suppose that a pressure of 10^8 dynes per square cm. diminishes the volume of a piece of steel by $5·55 \times 10^{-5}$ cubic cm. per cubic cm. The compression is therefore $5·55 \times 10^{-5}$, and hence the bulk modulus is

$$k = 10^8/(5·55 \times 10^{-5}) = 1·802 \times 10^{12} \text{ dynes per square cm.}$$

The only stress a liquid or a gas can permanently sustain is a hydrostatic pressure, which in the case of liquids under certain conditions may be *negative*[*], and hence the bulk modulus is the only elastic constant for a liquid or a gas. For this reason it is often spoken of as *the* elasticity of the liquid or the gas.

If the compression be not proportional to the pressure, we can still speak of the bulk modulus, but we then define it as the ratio of an infinitesimal increment of pressure from p to $p + dp$ to $- dv/v$, the corresponding diminution of volume per unit volume, the volume v being that which the body has under the pressure p. Hence we have, in the general case,

$$k = \frac{dp}{- dv/v} = - v \frac{dp}{dv} . \qquad \ldots\ldots\ldots\ldots\ldots(2)$$

The negative sign occurs since dv stands for the *increment* of volume corresponding to dp.

14. Rigidity. To produce a shear in a solid substance an appropriate stress is required. Let $ABA'B'$ (Fig. 3) be a cube of edge h, formed of elastic material, and let a uniform tangential stress of p dynes per square cm. be applied to the face A in a direction perpendicular to the line of intersection of A and B. The force ph^2 acting on A would cause a longitudinal acceleration of the

[*] See Poynting and Thomson, *Properties of Matter*, Chapter XL.

block unless resisted by an equal and opposite force.   If the balancing force were applied to A the block would not be strained. We therefore suppose that a uniform tangential stress of p dynes per square cm. is applied to the face A' in a direction opposite to

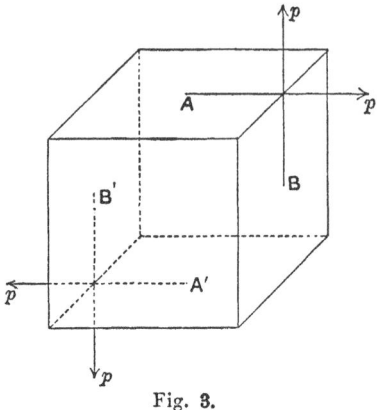

Fig. 3.

the stress on A.   The two forces, each equal to ph^2, which act on A and A', constitute a couple of moment ph^3 and would give the block an angular accleration about an axis parallel to the line of intersection of A and B, unless opposed by an equal and opposite couple.   This couple is supplied by uniform tangential stresses of p dynes per square cm. applied to the faces B and B' in the manner indicated in Fig. 3.   The remaining faces of the cube are not subjected to any forces.

The forces applied to the faces A, A', B, B' are in equilibrium and will cause no longitudinal or angular acceleration of the block.   But the forces will strain the block and will change the faces perpendicular to both A and B from squares into parallelograms with equal sides, as shown in Fig. 4.   Let the angles of each of these faces, when the block is strained, be $\frac{1}{2}\pi + \theta$ and $\frac{1}{2}\pi - \theta$ radians.

When the stress is small enough, it may be expected to be proportional to the shear θ, and experiment shows that Hooke's law

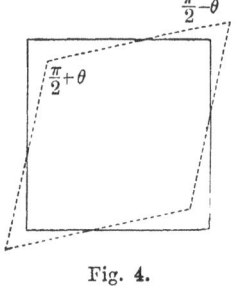

Fig. 4.

does express the relation between the stress and the shear when they are small. Thus, the ratio of the tangential stress on each of the faces A, B, A', B' to the resulting shear may be regarded as an "elastic constant" of the substance.

The ratio of the tangential stress to the shear is called the *rigidity* of the substance and is denoted by n. Since the shear is a pure number, the rigidity is measured in the same units as the stress, i.e. in dynes per square cm. Thus,

$$n = \text{rigidity} = \frac{\text{stress}}{\text{strain}} = \frac{p}{\theta}. \quad \ldots\ldots\ldots\ldots\ldots(3)$$

The rigidity is evidently the tangential stress which would produce a unit shear, i.e. a shear of one radian or $57° 18'$, if Hooke's law held for so great a strain.

As an example of the use of this formula, suppose that a tangential stress of 10^8 dynes per square cm. causes a shear of $1·22 \times 10^{-4}$ radians in steel. Then the rigidity is given by

$$n = \frac{10^8}{1·22 \times 10^{-4}} = 8·2 \times 10^{11} \text{ dynes per square cm.}$$

The quantity n is often called the *modulus of torsion**, because it makes its appearance in calculations respecting the torsion of a wire. But the term *modulus of torsion* is sometimes also used to denote the couple required to give a wire a twist of one radian per cm. of length. To avoid confusion, the term will not be used in this book.

If the shear be not proportional to the shearing stress, we can still speak of the rigidity of the substance, but we then define it as the ratio of an infinitesimal increase of shearing stress to the corresponding increase of the angle of shear. Hence in the general case,

$$n = \frac{dp}{d\theta}. \quad \ldots\ldots\ldots\ldots\ldots\ldots\ldots\ldots\ldots(4)$$

15. Stresses on the diagonal planes of a sheared cube. If we take a plane cutting the cube $ABA'B'$ (Fig. 3) and parallel either to the face A or to the face B, the stress is tangential to this plane and of amount p dynes per square cm., since the

---

* This term is used in Kohlrausch's *Introduction to Physical Measurements*, Third English Edition, p. 137.

uniformity of the strain demands that the stress on any plane
parallel to A should be equal to that on A and that the stress on
any plane parallel to B should be equal to that on B.

But now take a diagonal plane passing through the line of
intersection of the faces A and B' and dividing the cube into two
parts. The force on each of the faces A and B is ph^2, and thus
the resultant of these two forces is $2ph^2 . \cos \pi/4$ or $ph^2 \sqrt{2}$ at right
angles to the diagonal plane in a direction tending to separate one
part of the cube from the other. The area of the diagonal plane
is $h^2\sqrt{2}$, and hence the stress across this plane is a tension at right
angles to the plane amounting to p dynes per square cm. Since
the strain is uniform, there is an equal stress across every plane
parallel to this diagonal plane.

In a similar way it follows that the stress across the diagonal
plane, which passes through the line of intersection of the faces
A and B, is a pressure at right angles to this plane amounting to
p dynes per square cm. Since the strain is uniform, there is an
equal stress across every plane parallel to this diagonal plane.

The effect of these stresses will be to stretch the cube in the
direction of the tension and to compress it by an equal amount in
the direction of the pressure. Since the shear is p/n radians, we
see, by § 12, that the elongation in the direction of the tension
and the contraction in the direction of the pressure are each
$p/2n$ cm. per cm.

There will be no change of length in the direction perpendicular to both the pressure and the tension, since, if the pressure
produce an elongation in that direction, the tension will produce
an equal contraction. Since the stretching and compression due
to the stresses on the diagonal planes are equal, it follows that,
for small strains, the volume of the cube is unchanged by the
strain.

16. Alternative method of producing a shear. It may
be inferred, from the results of § 15, that a uniform shear can be
produced in a cubical block by a normal pressure of p dynes per
square cm. applied to one pair of faces a, a' while a normal tension
of p dynes per square cm. is applied to another pair b, b', the
remaining pair of faces being free from force. This distribution

of forces is shown in Fig. 5, where the arrow through the centre of any face represents the direction of the forces applied to that face. If each edge of the cube be h cm., the resultant of the forces acting on the faces a, b is $2ph^2 . \cos \pi/4$, or $ph^2\sqrt{2}$, parallel to the

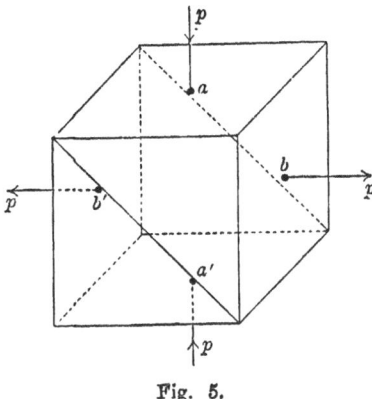

Fig. 5.

diagonal plane indicated in the figure. The area of this plane is $h^2\sqrt{2}$, and hence the stress on this plane is a shearing stress of p dynes per square cm.; a similar result holds for the other diagonal plane.

The material will, therefore, suffer a shear of p/n radians and it follows from § 12 that the lines in the cube, which are normal to the faces a, a', receive a contraction of $p/2n$ cm. per cm., and those which are normal to b, b' receive an elongation of equal amount, while those which are parallel to the line of intersection of the faces a and b are unchanged in length.

It will be noticed that the faces a, b of the cube of Fig. 5 correspond to the diagonal planes of the cube of Fig. 3, the stress in each case being a normal one. Further the faces A, B of Fig. 3 correspond to the diagonal planes of Fig. 5, the stresses being now tangential.

17. Young's modulus. When an evenly distributed pull of T dynes per square cm. is applied to each end of a straight uniform rod, the stress across any plane perpendicular to the axis of the rod is a uniform tension of T dynes per square cm.

The increase in the length of the rod, caused by this stress, is found by experiment to be proportional to the length of the rod, and, for small strains, to the tension, as Hooke's law leads us to expect.

The ratio of the longitudinal stress to the elongation, i.e. the increase of length per unit length, is called *Young's modulus* of the substance and is denoted by E; the longitudinal stress is to be calculated by dividing the total pull by the cross-section of the stretched rod. Since the elongation is a pure number, Young's modulus is measured in the same units as the stress, i.e. in dynes per square cm.

If a longitudinal stress of T dynes per square cm. increase the length of a rod from l cm. to $l + \lambda$ cm., the elongation, e, is λ/l cm. per cm. and hence

$$E = \text{Young's modulus} = \frac{\text{stress}}{\text{elongation}} = \frac{T}{e} = \frac{T}{\lambda/l} = \frac{Tl}{\lambda} \cdot \ldots (5)$$

As an example of the use of this formula, suppose that a total pull of 4×10^6 dynes applied to a steel wire 500 cm. in length and 5×10^{-2} cm. in radius, increases its length by 0·12 cm. The cross-section is $\pi \times 25 \times 10^{-4}$ or $7·85 \times 10^{-3}$ square cm. and thus the stress is $T = 4 \times 10^9/7·85$ or $5·10 \times 10^8$ dynes per square cm. The elongation is $e = 0·12/500$ or $2·4 \times 10^{-4}$ cm. per cm. Thus,

$$E = T/e = 5·1 \times 10^{12}/2·4 = 2·12 \times 10^{12} \text{ dynes per square cm.}$$

If the elongation be not proportional to the longitudinal stress, we can still speak of Young's modulus, but we then define it as the ratio of an infinitesimal increase of longitudinal stress dT to the corresponding elongation dl/l, where l is the length of the rod under the stress T. Hence, in the general case,

$$E = l \frac{dT}{dl} . \ldots \ldots \ldots \ldots \ldots (6)$$

Young's modulus, E, is not, in reality, a new elastic constant, since, in the case of an isotropic substance, we can show, as in § 19, that E is connected with the bulk modulus k and the rigidity n by the equation

$$\frac{1}{E} = \frac{1}{9k} + \frac{1}{3n}.$$

In the case of metals, Hooke's law fails, if the elongation much exceed $\frac{1}{1000}$ cm. per cm., because the rod receives a "permanent set" and does not return to its original length on the removal of the forces, and hence the greatest strains which can be employed in experiments are very small. When the rod is stretched, its cross section A becomes slightly less than A_0, the cross section of the unstretched rod, but the difference is so small that for practical purposes it is sufficient to calculate the stress by dividing the pull P by A_0. In fact, the experimental difficulties are such that it would be almost impossible to decide whether $A\lambda$ or $A_0\lambda$ is the more nearly proportional to the pull P.

18. Poisson's ratio. When a rod or wire is stretched by forces applied to its ends, while its sides are free from force, it is found that its cross-section diminishes, and hence the strain is not a simple elongation but an elongation accompanied by a contraction in every direction perpendicular to the elongation. For small elongations the ratio of the contraction to the elongation is constant for a given specimen, but the ratio varies from substance to substance.

Let the elongation of the rod, i.e. the increase of length per unit length, parallel to its axis be e cm. per cm., and let the lateral contraction, i.e. the diminution of length per unit length, of lines at right angles to the axis be f cm. per cm. Then the ratio of f to e is called *Poisson's ratio* and is denoted by σ. Thus

$$\sigma = \text{Poisson's ratio} = \frac{\text{Lateral contraction}}{\text{Elongation}} = \frac{f}{e}. \quad \dots(7)$$

Since both the elongation and the contraction are pure numbers, Poisson's ratio is a pure number and is independent both of the unit of length and of the unit of force.

As an example of the use of this formula, suppose that, when a steel wire 1000 cm. in length and 0·1 cm. in diameter is stretched by 0·4 cm., its diameter diminishes by $1·12 \times 10^{-5}$ cm. Then the elongation is $0·4/1000$ or 4×10^{-4}, and the lateral contraction is $1·12 \times 10^{-5}/0·1$ or $1·12 \times 10^{-4}$. Hence

$$\sigma = \text{Poisson's ratio} = \frac{\text{Lateral contraction}}{\text{Elongation}} = \frac{1·12 \times 10^{-4}}{4 \times 10^{-4}} = ·28.$$

If a rod of length l, having for its section a square of side a, receive the elongation e, each side suffers the contraction f, and thus after the strain the length is $l(1 + e)$ and the cross-section is $a^2(1 - f)^2$ or $a^2(1 - 2f)$, because f is very small. Since $f = e\sigma$, we see that the elongation e is accompanied by a diminution of cross-section of $2\sigma e$ square cm. per square cm., and by an increase of volume of $(1 + e)(1 - 2\sigma e) - 1$ c.c. per c.c. or $e(1 - 2\sigma)$ c.c. per c.c., when e^2 is neglected.

In an actual experiment on a metal wire it would be difficult, by any simple means, to make a direct measurement of the contraction, and hence σ is generally found by some indirect method.

Poisson's ratio σ is not an independent elastic constant, since, in the case of an isotropic substance, we can show, as in § 19, that σ is connected with the bulk modulus k and the rigidity n by the equation

$$\sigma = \frac{3k - 2n}{6k + 2n}.$$

19. Relations between elastic constants. In the case of an isotropic substance, two mathematical relations connect the bulk modulus k, the rigidity n, Young's modulus E, and Poisson's

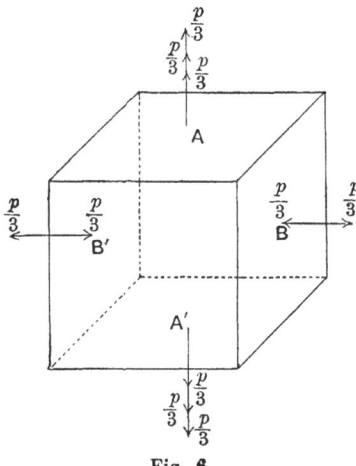

Fig. 6.

ratio σ, and thus only two out of these four quantities are independent. The two relations may be found in the following manner:—Let A, A', B, B', C, C' (Fig. 6) denote the six faces of a cube of edge h, and let a uniform normal tension of p dynes per square cm. be applied to the faces A and A'. By § 17, these forces will produce an elongation of p/E cm. per cm. parallel to the tension, and, by § 18, a contraction at right angles to B of $\sigma p/E$ cm. per cm. and an equal contraction at right angles to C.

But the same effect can be produced in another way. Replace the tension p on the faces A, A' by three superposed tensions each equal to $\frac{1}{3}p$, and apply to each of the faces B, B', C, C' a tension of $\frac{1}{3}p$ and a pressure of $\frac{1}{3}p$, as indicated in Fig. 6, where each arrow head stands for $\frac{1}{3}p$. To avoid confusion, the forces acting on the faces C, C' are not shown in Fig. 6. The pressure then exactly neutralises the tension on each of these four faces.

The tensions of $\frac{1}{3}p$ on the six faces are equivalent to a hydrostatic pressure of $-\frac{1}{3}p$, and hence, by § 13, cause a uniform expansion of $p/3k$ c.c. per c.c., increasing the volume of the cube from h^3 to $h^3(1 + p/3k)$ and the edges of the cube from h to $h(1 + p/3k)^{\frac{1}{3}}$. Expanding by the binomial theorem and rejecting p^2/k^2 and higher powers of p/k, we find that the edges are increased to $h(1 + p/9k)$. Hence the tensions of $\frac{1}{3}p$ on the six faces cause an elongation of $p/9k$ in every direction.

We must now take account of the pressures $\frac{1}{3}p$ on the faces B, B', C, C', and we begin by considering the pressures on B and B' in conjunction with one pair of the three partial tensions $\frac{1}{3}p$ on the faces A and A'. By § 16, this set of forces will cause an elongation $\frac{1}{3}p/2n$ or $p/6n$ cm. per cm. at right angles to A, a contraction $p/6n$ cm. per cm. at right angles to B, but no change of length at right angles to C. In the same way, the pressures $\frac{1}{3}p$ on the faces C, C', taken in conjunction with the remaining pair of partial tensions on the faces A, A', will cause an elongation $p/6n$ at right angles to A, a contraction $p/6n$ at right angles to C, and no change of length at right angles to B.

Collecting these results, we find that the resultant elongation in the direction of the original tension is

$$\frac{p}{9k} + \frac{p}{6n} + \frac{p}{6n} = p\left(\frac{1}{9k} + \frac{1}{3n}\right) \text{ cm. per cm.}$$

But we have seen in § 17 that the elongation, when expressed in terms of Young's modulus, is p/E cm. per cm. Thus, by equating the two expressions for the elongation, we have

$$\frac{1}{E} = \frac{1}{9k} + \frac{1}{3n}. \quad \dots\dots\dots\dots\dots\dots(8)$$

The resultant contraction at right angles to either B or C or in any direction at right angles to the original tension is

$$\frac{p}{6n} - \frac{p}{9k} \text{ cm. per cm.}$$

But, by § 18, this contraction, when expressed in terms of Poisson's ratio and Young's modulus, is σe or $\sigma p/E$ cm. per cm. Equating the two expressions for the contraction, we have

$$\frac{\sigma}{E} = \frac{1}{6n} - \frac{1}{9k}. \quad \dots\dots\dots\dots\dots\dots(9)$$

Hence, by (8),

$$\sigma = \frac{3k - 2n}{6k + 2n}. \quad \dots\dots\dots\dots\dots\dots(10)$$

By equations (8) and (9), when we know the values of any two of the four quantities k, n, E and σ, we can calculate the values of the other two. The two which are usually found by experiment are Young's modulus and the rigidity.

If we add (8) to (9), we easily find

$$\sigma = \frac{E}{2n} - 1. \quad \dots\dots\dots\dots\dots\dots(11)$$

If we eliminate n between (8) and (9), we have

$$\sigma = \frac{1}{2} - \frac{E}{6k}.$$

From (10) we find

$$3k\,(1 - 2\sigma) = 2n\,(1 + \sigma). \quad \dots\dots\dots\dots(12)$$

Hence, if 2σ were greater than 1 or if σ were less than -1, either k or n would be negative. It therefore follows that, for an isotropic solid, Poisson's ratio cannot exceed $\frac{1}{2}$ and cannot be less than -1.

20. Isothermal and adiabatic elasticities. When the form of a body is changed by the application of forces, there is, as a rule, a rise or fall of temperature. This effect is very conspicuous in gases; it may also be easily observed in the case of india-rubber. Thus, if an india-rubber band be suddenly stretched, there will be a rise of temperature which may be easily detected by bringing the stretched band into contact with the lips.

If the temperature of each part of the body be maintained constant while the forces do their work upon the body, there will be a definite relation between the forces and the changes of form which they produce. The elastic constants corresponding to this isothermal condition, will be denoted by k_t, E_t and n_t, the subscript t denoting that the temperature is constant.

On the other hand, if no heat be allowed to enter or leave any part of the body, the temperature will change in a definite manner corresponding to the action of the forces, and, since the relation between the forces and the changes of form depends upon the temperature, this relation, though still definite, will differ from that which holds when the isothermal condition is satisfied. When no heat enters or leaves any part of the body, the change of form is said to take place under the adiabatic condition. Now if dQ be the heat which enters a perfectly elastic body under any conditions when the forces receive any given small increments, an equal amount of heat will be given out under the same conditions when the forces return to their initial values. Thus the change is a reversible one and hence, by the principles of thermodynamics*, we can write

$$dQ = td\phi, \quad \dots\dots\dots\dots\dots\dots(13)$$

where $d\phi$ is the increase of entropy corresponding to the increase of heat dQ and t is the temperature measured from the absolute zero. Hence the elastic constants corresponding to the adiabatic condition will be denoted by k_ϕ, E_ϕ and n_ϕ, the subscript ϕ denoting that the entropy is constant. We may also denote the adiabatic elastic constants by k_Q, E_Q and n_Q when it is convenient to do so, the subscript Q denoting that no heat enters or leaves any part of the body.

In some of the experiments described below, statical effects are observed, as when a wire is stretched by a load. Here we may suppose that any change of temperature, due to the application

---

* See Maxwell, *Theory of Heat*, Chapter VIII, or Poynting and Thomson, *Heat*, Chapter XVII.

of the load, has disappeared by radiation and by conduction to the surrounding air before the extension is observed. In other experiments dynamical methods are employed and the vibrations of the system are observed. If the vibrations were infinitely rapid, adiabatic conditions would prevail, since there would be no time for radiation or conduction to cause any appreciable transfer of heat. In practice the time of vibration is finite, and thus there will be some departure from adiabatic conditions. We shall be able to estimate how close an agreement may be expected between the results of statical methods and those of dynamical methods, if we know the relation between the isothermal and the adiabatic elasticities, for it is easily seen that the value of any modulus found by a dynamical method, when the time of vibration is finite, will lie between the isothermal and the adiabatic values of that modulus.

21. Ratio of adiabatic to isothermal elasticity. In discussing the applications of thermodynamics to elasticity, it is convenient to express the moduli of elasticity in terms of differential coefficients. In the case of the bulk modulus we have, by equation (2), § 13,

$$k_\phi = k_Q = -v \left(\frac{dp}{dv}\right)_\phi = -v \left(\frac{dp}{dv}\right)_Q, \quad \ldots\ldots\ldots\ldots(14)$$

$$k_t = -v \left(\frac{dp}{dv}\right)_t. \quad \ldots\ldots\ldots\ldots\ldots\ldots(15)$$

The subscript Q denotes that the variations p and v are so related that Q does not change, i.e. so that no heat enters or leaves the substance. Similarly the subscript t denotes that the temperature is constant.

Now, if z be a function of two independent variables x and y, we have

$$dz = \left(\frac{dz}{dx}\right)_y dx + \left(\frac{dz}{dy}\right)_x dy.$$

If the variations in x and y cause no resultant change in z, there must be a definite relation between dx and dy. Putting $dz = 0$, we find the relation to be

$$\left(\frac{dy}{dx}\right)_z = -\left(\frac{dz}{dx}\right)_y \Big/ \left(\frac{dz}{dy}\right)_x = -\left(\frac{dz}{dx}\right)_y \left(\frac{dy}{dz}\right)_x.$$

Since the state of a perfectly elastic homogeneous body, which is subjected to a uniform hydrostatic pressure p, is completely defined when p and v are known, t is a function of the two independent variables p and v, and hence, if p and v be so related that t does not vary,

$$\left(\frac{dp}{dv}\right)_t = -\left(\frac{dt}{dv}\right)_p \left(\frac{dp}{dt}\right)_v = -\left(\frac{dp}{dt}\right)_v \Big/ \left(\frac{dv}{dt}\right)_p .$$

If, however, p and v be so related that Q does not vary,

$$\left(\frac{dp}{dv}\right)_Q = -\left(\frac{dQ}{dv}\right)_p \left(\frac{dp}{dQ}\right)_v = -\left(\frac{dQ}{dv}\right)_p \Big/ \left(\frac{dQ}{dp}\right)_v ,$$

and hence, by (14) and (15),

$$\frac{k_\phi}{k_t} = \frac{\left(\dfrac{dp}{dv}\right)_Q}{\left(\dfrac{dp}{dv}\right)_t} = \frac{\left(\dfrac{dQ}{dv}\right)_p \left(\dfrac{dv}{dt}\right)_p}{\left(\dfrac{dQ}{dp}\right)_v \left(\dfrac{dp}{dt}\right)_v} = \frac{\left(\dfrac{dQ}{dt}\right)_p}{\left(\dfrac{dQ}{dt}\right)_v} .$$

If the mass of the body be m,

$$\left(\frac{dQ}{dt}\right)_p = mC_p, \qquad\qquad \left(\frac{dQ}{dt}\right)_v = mC_v,$$

where C_p and C_v are the specific heats at constant pressure and at constant volume respectively. Hence

$$\frac{k_\phi}{k_t} = \frac{C_p}{C_v}.$$

The reader must bear in mind that the specific heat of a substance, under any given condition, is the amount of *heat* which is absorbed when one gramme of the substance rises from $t°$ to $(t+1)°$ under the given condition, and must remember that, in estimating the specific heat, we entirely leave out of account any energy which may be supplied mechanically, as by the action of a pressure when the volume diminishes.

In the case of Young's modulus, the state of a given wire is defined when we know its length and the tension, i.e. the force per unit area acting across a transverse section, and thus a process similar to that adopted for k will give

$$\frac{E_\phi}{E_t} = \frac{C_T}{C_l},$$

where C_T is the specific heat for constant tension and C_l the specific heat for constant length.

In the case of the rigidity, we have in like manner

$$\frac{n_\phi}{n_t} = \frac{C_p}{C_\theta},$$

where C_p is the specific heat when p, the shearing stress, is constant, and C_θ is the specific heat for constant angle of shear. If the shear have the constant value zero, the shearing stress remains zero in spite of a change of temperature, for no shearing stress is required to maintain the cubical form of a cubical block when its temperature is changed. Hence the heat absorbed while the temperature rises $1°$ when the shearing stress is zero is equal to that absorbed during the same rise of temperature when the shear is zero, for the two conditions are identical. Thus C_p for zero stress is identical with C_θ for zero shear. Hence for infinitesimal shears we may write

$$n_\phi = n_t.$$

In practical determinations of the rigidity, we take care that the shear θ is always very small (perhaps not exceeding $\frac{1}{1000}$ radian) in order to keep within the limits of Hooke's law. We may therefore conclude that, in such determinations, the same value of the rigidity will be obtained whether n be found by a statical method, where the conditions are isothermal, or by a dynamical method, where the conditions are more or less nearly adiabatic.

Returning to the bulk modulus, we may expect that C_p will differ from C_v, and that, in consequence, k_ϕ will differ from k_t. When the pressure remains constant while the temperature rises from t to $t + dt$, the volume will generally change, and then external work will be done. On the other hand, when the volume remains constant while the temperature rises from t to $t + dt$, the pressure will generally change, but no external work will be done, and thus the final conditions of the body are different in the two cases. Hence we cannot say that equal amounts of heat are absorbed in the two cases, even though, as may be the case with an elastic solid, the initial pressure be zero.

Similar considerations apply to Young's modulus, and thus we may expect that C_T, the specific heat for constant tension, will generally differ from C_l, the specific heat for constant length.

In the exceptional case where the substance, when subjected to a constant pressure p, has a point of maximum density at the temperature t, an infinitesimal change of temperature from t to $t + dt$ will cause no change of volume, provided the pressure remain constant, and thus the two conditions of constant volume and constant pressure become indistinguishable, and C_p becomes identical with C_v.

The experimental comparison of C_p with C_v or of C_T with C_l would probably be more difficult than the direct experimental comparison of k_ϕ with k_t or of E_ϕ with E_t, and thus we must look for some other method of finding a relation between the adiabatic and the isothermal values for each of these two moduli.

22. Difference between reciprocals of isothermal and adiabatic elasticities. Though it is impracticable to calculate the *ratio* of the adiabatic to the isothermal value of k or of E from the ratio of the specific heat for constant stress to that for constant strain, yet we can find the *difference* between the reciprocals of these values in terms of quantities which can be determined.

We shall now find the difference in the case of the bulk modulus. On the p-v diagram for unit mass of a substance let AB (Fig. 7) be an isothermal line and AC an adiabatic line or line of constant entropy. Let the pressure and the volume of the unit mass at A be p, v and let the corresponding temperature be t. Let the line of constant pressure $p - dp$ cut the isothermal and adiabatic lines through A in B and C and the line of constant volume v in D,

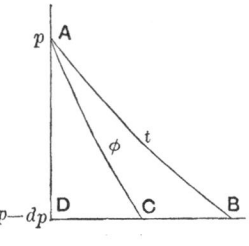

Fig. 7.

and let the temperature at C be $t - dt$. Now DB is the increase of volume which occurs when the pressure *falls* by dp while the temperature remains constant, and DC corresponds in a similar way to constant entropy. Hence

$$DB = - (dv/dp)_t\, dp, \qquad DC = - (dv/dp)_\phi\, dp.$$

Further, CB is the increase of volume which occurs when the temperature rises by dt while the pressure remains constant, and thus

$$CB = (dv/dt)_p \, dt.$$

But $$DB - DC = CB,$$

and hence, by (14) and (15), we have

$$v \left(\frac{1}{k_t} - \frac{1}{k_\phi} \right) dp = \left(\frac{dv}{dt} \right)_p dt.$$

Here dt is the rise of temperature which occurs when the pressure rises by dp, while the entropy remains constant, and hence

$$dt = (dt/dp)_\phi \, dp.$$

But, by Maxwell's second thermodynamic relation*,

$$(dt/dp)_\phi = (dv/d\phi)_p, \quad \dots\dots\dots\dots\dots\dots(16)$$

and thus, since

$$\left(\frac{dv}{d\phi} \right)_p = \left(\frac{dv}{dt} \right)_p \left(\frac{dt}{d\phi} \right)_p,$$

we have

$$\frac{1}{k_t} - \frac{1}{k_\phi} = \frac{1}{v} \left(\frac{dv}{dt} \right)_p \left(\frac{dv}{d\phi} \right)_p = \frac{1}{v} \left(\frac{dv}{dt} \right)_p^2 \left(\frac{dt}{d\phi} \right)_p.$$

But, by (13), since we have unit mass,

$$t \, (d\phi/dt)_p = (dQ/dt)_p = C_p,$$

and hence, finally,

$$\frac{1}{k_t} - \frac{1}{k_\phi} = \frac{t}{vC_p} \left(\frac{dv}{dt} \right)_p^2 = \frac{tv}{C_p} \left(\frac{1}{v} \frac{dv}{dt} \right)_p^2. \quad \dots\dots\dots (17)$$

From this equation we can find the difference $k_t^{-1} - k_\phi^{-1}$, when we know the absolute temperature (t), the volume (v) of unit mass, the specific heat at constant pressure (C_p), expressed in mechanical units, and $(v^{-1} dv/dt)_p$, the coefficient of cubical expansion under the constant pressure p.

Since the right side of (17) cannot be negative, it follows that k_ϕ is greater than k_t unless the substance be at a point of maximum or minimum density, like water at 4° C., when the coefficient of cubical expansion under constant pressure, $(v^{-1} dv/dt)_p$, vanishes. When this is the case, k_ϕ is equal to k_t.

---

* See Maxwell, *Theory of Heat*, Chapter IX, or Tait, *Heat*, Chapter XXI, or Preston, *Theory of Heat*, Chapter VIII, Section IV.

In the case of copper at $0°$ C., when $t = 273$ on the absolute scale, we have approximately

$v = 0\cdot11$ c.c., $\qquad (v^{-1}\, dv/dt)_p = 5 \times 10^{-5}$ degree^{-1},

$C_p = 0\cdot095 \times 4\cdot2 \times 10^7 = 4 \times 10^6$ ergs per grm. per deg.

$k_t = 1\cdot7 \times 10^{12}$ dyne cm.$^{-2}$.

Hence, by (17),

$$\frac{k_t}{k_\phi} = 1 - \frac{tvk_t}{C_p}\left(\frac{1}{v}\frac{dv}{dt}\right)_p^2 = 1 - \frac{273 \times 0\cdot11 \times 1\cdot7 \times 10^{12}}{4 \times 10^6}(5 \times 10^{-5})^2$$

$$= 1 - 0\cdot032.$$

Thus k_ϕ is about 3 per cent. greater than k_t.

The process employed for the bulk modulus can be applied, with slight changes, to give the difference between the reciprocals of the adiabatic and the isothermal values of Young's modulus. If the length of a rod of unit mass and of unit cross-section be l cm., the work which the *rod* does when l increases by dl cm., is $- Tdl$ ergs, where T dyne cm.$^{-2}$ is the tensile stress in the rod. Comparing this with pdv, the work done by a body in expanding against a pressure p, we see that, if we write $- dT$ for dp and dl for dv, in Maxwell's equation (16), the resulting equation

$$(dt/dT)_\phi = - (dl/d\phi)_T \quad \ldots\ldots\ldots\ldots\ldots(18)$$

will apply to the stretching of a rod of unit mass and unit section by a tensile stress T.

We can apply Fig. 7 to this case by measuring T in the direction DA and l in the direction DB. Using the definition of Young's modulus given by equation (6), § 17, we then obtain,

$$DB = - (dl/dT)_t\, dT = - (l/E_t)\, dT,$$
$$DC = - (dl/dT)_\phi\, dT = - (l/E_\phi)\, dT,$$
$$CB = (dl/dt)_T\, dt.$$

But $DB - DC = CB$, and hence

$$l\left(\frac{1}{E_t} - \frac{1}{E_\phi}\right) dT = -\left(\frac{dl}{dt}\right)_T dt.$$

Here dt is the rise of temperature which occurs when the tension rises by dT, while the entropy remains constant. Hence, by (18),

$$dt = (dt/dT)_\phi\, dT = - (dl/d\phi)_T\, dT,$$

and thus, since

$$\left(\frac{dl}{d\phi}\right)_T = \left(\frac{dl}{dt}\right)_T \left(\frac{dt}{d\phi}\right)_T,$$

we have

$$\frac{1}{E_t} - \frac{1}{E_\phi} = \frac{1}{l}\left(\frac{dl}{dt}\right)_T \left(\frac{dl}{d\phi}\right)_T = \frac{1}{l}\left(\frac{dl}{dt}\right)_T^2 \left(\frac{dt}{d\phi}\right)_T.$$

But, by (13), since we have unit mass,

$$t\,(d\phi/dt)_T = (dQ/dt)_T = C_T,$$

the specific heat under constant tension, and hence finally

$$\frac{1}{E_t} - \frac{1}{E_\phi} = \frac{t}{lC_T}\left(\frac{dl}{dt}\right)_T^2 = \frac{tl}{C_T}\left(\frac{1}{l}\frac{dl}{dt}\right)_T^2. \quad \dots\dots\dots(19)$$

Hence E_ϕ is greater than E_t unless $(l^{-1}dl/dt)_T$, the coefficient of linear expansion under constant tension, be zero.

When there is zero stress, the coefficient of linear expansion is one-third of that of cubical expansion and also C_T is equal to C_p, and thus, since l in (19) is numerically equal to v in (17), it follows that, when the stresses are infinitesimal, the right side of (19) is one-ninth of the right side of (17). Thus

$$\frac{1}{E_t} - \frac{1}{E_\phi} = \frac{1}{9}\left(\frac{1}{k_t} - \frac{1}{k_\phi}\right)$$

or

$$1 - \frac{E_t}{E_\phi} = \frac{E_t}{9k_t}\left(1 - \frac{k_t}{k_\phi}\right).$$

But, by § 19, $E_t/k_t = 3 - 6\sigma$, and thus

$$1 - \frac{E_t}{E_\phi} = \tfrac{1}{3}(1 - 2\sigma)\left(1 - \frac{k_t}{k_\phi}\right).$$

When, as in the case of metals, σ is about $\tfrac{1}{4}$,

$$1 - \frac{E_t}{E_\phi} = \frac{1}{6}\left(1 - \frac{k_t}{k_\phi}\right)$$

and thus E_t/E_ϕ is much more nearly unity than k_t/k_ϕ.

In the case of copper at $0°$ C., when $t = 273$, we have approximately

$$l = 0.11 \text{ cm.} \qquad (l^{-1}\,dl/dt)_T = 1.7 \times 10^{-5} \text{ degree}^{-1},$$
$$C_T = 0.095 \times 4.2 \times 10^7 = 4 \times 10^6 \text{ ergs per grm. per deg.}$$
$$E_t = 1.2 \times 10^{12} \text{ dyne cm.}^{-2}.$$

Hence, by (19),

$$\frac{E_t}{E_\phi} = 1 - \frac{tlE_t}{C_T}\left(\frac{1}{l}\frac{dl}{dt}\right)^2_T = 1 - \frac{273 \times 0{\cdot}11 \times 1{\cdot}2 \times 10^{12}}{4 \times 10^6}(1{\cdot}7 \times 10^{-5})^2$$
$$= 1 - 0{\cdot}0026.$$

Thus, E_ϕ is only about three parts in a thousand greater than E_t.

In most cases the experimental difficulties make it impossible to measure either E_t or E_ϕ to within one per cent., and hence for most purposes we may disregard the distinction between E_t and E_ϕ.

We could find the difference between the reciprocals of the adiabatic and the isothermal values of the rigidity by a process similar to that employed for k and E, but it will be more interesting to deduce the result for n from those obtained for k and E by aid of the relation connecting n with k and E. Thus, by equation (8), § 19,

$$\frac{1}{n} = \frac{3}{E} - \frac{1}{3k},$$

and hence, by (17) and (19),

$$\frac{1}{n_t} - \frac{1}{n_\phi} = \frac{3tl}{C_T}\left(\frac{1}{l}\frac{dl}{dt}\right)^2_T - \frac{tv}{3C_p}\left(\frac{1}{v}\frac{dv}{dt}\right)^2_p. \quad\dotsc\dotsc\dotsc\dotsc(20)$$

If we consider only infinitesimal strains in a substance initially free from strain, we may put

$$C_T = C_p, \qquad (v^{-1}\,dv/dt)_p = 3\,(l^{-1}\,dl/dt)_T,$$

since the coefficient of cubical expansion is three times that of linear expansion. Since l is the length of a rod of unit mass and unit cross-section, it is numerically equal to v, the volume of unit mass. Hence the right side of (20) is zero, and thus

$$n_\phi = n_t,$$

the same result as that found in § 21.

CHAPTER II.

23. Practical applications of the theory of elasticity.
In Chapter I the elementary principles of elasticity have been explained, and in Chapter III some experimental methods of determining the elastic constants will be given. Before we pass on to those applications of the theory of elasticity, it will be convenient to give, in the present chapter, the mathematical solutions of some important problems, since these solutions are required in connexion with several experiments. The remaining problems will be considered as they arise in the course of the experimental work.

Theoretical elasticity suffers from the disadvantage that *exact* mathematical solutions for finite strains have been obtained in very few problems, most of the investigations given in text books on the subject depending upon the assumption that the strains are infinitesimal. In most cases it is further assumed that the stresses which act on the faces of any element of volume, when an elastic body is strained, produce the same effects as if the element had continued to occupy its *original* position. This assumption is justifiable in many instances, but it sometimes leads to results which become erroneous when the strains cease to be infinitesimal, though they may remain very small—so small, in fact, that there is no question as to the possible failure of Hooke's law.

In this chapter we shall endeavour to indicate the points at which assumptions are made and to consider the difficulties which arise. The investigations will perhaps appear rather lengthy, but it is hoped that this will not be thought a serious disadvantage.

In practical work in elasticity, additional difficulties make their appearance. Thus, it is seldom, if ever, possible to apply to the surface of the body the distribution of stress corresponding to the mathematical solutions given in this chapter, and hence it often becomes necessary to rely more on the light of instinct, instructed and guided by quantitative experiments, than on strict mathematical analysis. In addition, there is the practical difficulty that we have no means of ascertaining to what extent a given rod or wire is non-isotropic. In some cases, indeed, the experimental results show that the theory of isotropic elastic solids, which is given in Chapter I, entirely fails to account for the experimental facts *.

24. Principle of Saint-Venant. In many practical cases it is impossible to produce exactly that distribution of stress over the surface of an elastic body, which is assumed for the purpose of obtaining a problem capable of solution by comparatively simple mathematics. In these cases we fall back on a principle stated by Barré de Saint-Venant in 1855. " According to this principle, the strains that are produced in a body by the application, to a small part of its surface, of a system of forces statically equivalent to zero force and zero couple, are of negligible magnitude at distances which are large compared with the linear dimensions of the part†."

In any given case, much depends upon what is meant by "strains of negligible magnitude " and by " distances which are large compared with the linear dimensions of the part." But, when the body takes the form of a rod, there is mathematical evidence to the effect that if S_1 and S_2 denote two different systems of forces, which, taken together, are statically equivalent to zero force and zero couple, and if S_1 and S_2 be simultaneously applied to the rod near one of its ends, the resulting strain at any point diminishes very rapidly as the distance of the point from the end increases and is much less than one hundredth of the strain due to either S_1 or S_2 acting alone, provided the distance of the point from the end exceeds twice the greatest width of the transverse section of the rod.

* See, for example, EXPERIMENT 8, Chapter III.

† A. E. H. Love, *Treatise on the Mathematical Theory of Elasticity*, Second Edition, p. 129.

The distribution of force over the ends of a rod, which the simple theory demands in any given case, cannot, as a rule, be produced in an actual experiment. But Saint-Venant's principle assures us that, if we apply to the rod, in the neighbourhood of an end, *any* distribution of force statically equivalent to the system which the simple theory assigns to the end of the rod, the state of stress and strain set up in the interior of the rod is practically the same as in the simple theory, except, of course, in the immediate neighbourhood of the end. An example of the application of Saint-Venant's principle is noticed in § 41.

In any particular problem, the principle may be tested by varying the length of the rod and comparing the results obtained for different lengths. Thus, in the Practical Example of Ex-PERIMENT 4, Chapter III, it was found that, within the error of experiment, the angle through which the pointer turned, when a given couple was applied to the rod, was proportional to the distance of the pointer from the face of the block into which the fixed end of the rod is soldered. In this experiment the rod is held by forces applied to the curved surface of the rod, while in the theory of § 39 the forces are applied to the plane end of the rod. We may conclude that, in the experiment, the strains at distances from the face of the block exceeding one centimetre did not differ appreciably from those which would have existed if the rod had ended in a plane at the face of the block and the forces discussed in § 39 had been applied to that plane.

25. Dr Filon's results for tension. Dr L. N. G. Filon[*] has obtained the necessary mathematical formulae and from them has deduced numerical results which enable us to gain some idea of the character of the strain in a circular cylinder under tension, when the pull is not applied evenly over the ends of the cylinder, but is produced by tangential forces acting on the sides of the cylinder. These results show that the strain at a point near the surface of the cylinder is practically independent of the manner in which the pull is applied, provided that the distance of the point

---

[*] "On the elastic equilibrium of circular cylinders under certain practical systems of load." *Phil. Trans. Royal Society*, Vol. 198, A, pp. 147—233.

from the nearest point of application of the pull exceeds half the radius of the cylinder.

Dr Filon considers the case of a cylinder AA' (Fig. 8) of length $2c$ and radius a. He takes

$$AB = A'B' = \tfrac{1}{3}c, \quad BC = B'C' = \tfrac{1}{3}c, \quad CO = C'O = \tfrac{1}{3}c,$$

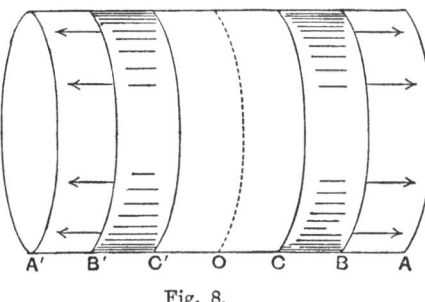

Fig. 8.

and supposes that tangential stresses parallel to the axis of the cylinder of amount p dynes per square cm. are applied to the bands BC and $B'C'$. The total pull is thus $\tfrac{2}{3}\pi acp$. The remainder of the surface of the cylinder is free from stress.

On account of symmetry, the molecules in the central transverse plane through O will not be displaced parallel to the axis of the cylinder when the pull is applied.

Denoting by w the increase of distance of a particle from the central plane, Dr Filon has calculated the ratio of w to w_0, where w_0 is the displacement which the end of the cylinder would have had if the pull $\tfrac{2}{3}\pi acp$ had been uniformly distributed over the plane ends of the cylinder. The original distance of the particle from the central plane is z and its original distance from the axis is r. The results in the table * have been calculated for the case in which Poisson's ratio (σ) is $\tfrac{1}{4}$ and the total length of the cylinder is half its circumference, so that $\pi a = 2c$.

All the particles originally at a distance of $c/10$ or $\pi a/20$ from the central plane would have had the displacement $w_0/10$ if the pull had been applied to the ends of the cylinder. From the table we see that at the surface, where $r = a$, w is $0 \cdot 1097 w_0$ or

* *Phil. Trans.* Vol. 198, A, p. 172.

about 10 per cent. greater than $w_0/10$, and this approximation to uniform stretching has been reached in a distance as small as about $a/3$ from the nearest point of application of the tangential force. If the distances CO and $C'O$ had been greater in proportion to the diameter of the rod than in Dr Filon's calculations, the displacements of points near the central plane would have differed still less from those occurring when the rod is uniformly stretched by a force $\frac{2}{3}\pi acp$ evenly applied to its plane ends.

Values of w/w_0.

z	$r=0$	$r=0\cdot2a$	$r=0\cdot4a$	$r=0\cdot6a$	$r=a$
0	0·0000	0·0000	0·0000	0·0000	0·0000
0·1c	0·0569	0·0600	0·0699	0·0868	0·1097
0·2c	0·1113	0·1169	0·1349	0·1675	0·2290
0·3c	0·1624	0·1695	0·1926	0·2368	0·3825
0·4c	0·2113	0·2192	0·2446	0·2949	0·5924
0·5c	0·2601	0·2683	0·2947	0·3471	0·6715
0·6c	0·3090	0·3174	0·3442	0·3954	0·6881
0·7c	0·3549	0·3636	0·3903	0·4372	0·5742
0·8c	0·3930	0·4016	0·4270	0·4668	0·5176
0·9c	0·4181	0·4265	0·4500	0·4825	0·4974
c	0·4268	0·4351	0·4577	0·4872	0·4920

The displacements of points on the surface are much more nearly equal to those in a uniformly stretched rod than are the displacements of points near the axis. Thus, for points on the axis, where $r=0$, the displacement of the point for which $z=c/10$ is only about $w_0/18$ instead of $w_0/10$. This was to be expected, since the chief part of the longitudinal pull is borne by the outer layers of the cylinder; the inner core, therefore, is comparatively little stretched. But the difference in the extension of the inner core and the outer layer is not of much consequence in experiments, since our observations are restricted to the surface of the cylinder.

The student will find it instructive to plot a curve for each value of z showing how w depends upon r.

In another paper*, Dr Filon has considered the case of an infinitely long rod XX' (Fig. 9) of rectangular section with sides $2a$ and $2b$ centimetres, the side $2b$ being either very great or very small compared with the side $2a$. To the end X a longitudinal force of $4abF$ dynes is evenly applied so that the tension at the

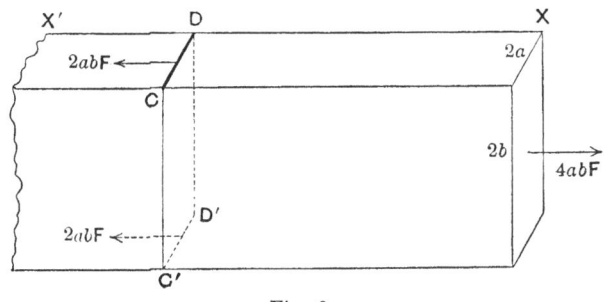

Fig. 9.

end is F dynes per square centimetre. At a great distance from X, a force of $2abF$ dynes is evenly applied to the surface of the rod along each of the lines CD and $C'D'$, in which a transverse plane cuts the rod. The force per unit length of each of these lines is thus bF dynes per centimetre. The lines CC' and DD' have no forces applied to them and the end X' is free from stress. Dr Filon has calculated the elongation e at a point P on the face containing CD at a distance x from CD, x being counted positive when P lies between CD and X. If the rod had been stretched by two forces each equal to $4abF$ evenly applied at X and X', the elongation would have had the constant value F/E, where E is Young's modulus (§ 17, Chapter I). The following table†, which

* "On an approximate solution for the bending of a beam of rectangular cross-section under any system of load." *Phil. Trans. Royal Society*, Vol. 201, A, pp. 63—155. Some changes have been made in Dr Filon's notation.

† *Phil. Trans.* Vol. 201, A, p. 145.

has been calculated for the case where Poisson's ratio is $\frac{1}{4}$, gives the ratio of e to F/E for a series of values of x.

x	$\dfrac{e}{F/E}$	x	$\dfrac{e}{F/E}$
πb	$+0\cdot997$	$-\frac{1}{6}\pi b$	$-0\cdot652$
$\frac{2}{3}\pi b$	$+0\cdot982$	$-\frac{1}{3}\pi b$	$-0\cdot084$
$\frac{1}{2}\pi b$	$+0\cdot985$	$-\frac{1}{2}\pi b$	$+0\cdot015$
$\frac{1}{3}\pi b$	$+1\cdot084$	$-\frac{2}{3}\pi b$	$+0\cdot018$
$\frac{1}{6}\pi b$	$+1\cdot652$	$-\pi b$	$+0\cdot003$

It will be seen that the surface elongation reaches its limiting value F/E with great rapidity. At a distance from the lines of application of the pull equal to πb, i.e. equal to about one and a half times the distance between those lines, the elongation differs from F/E by only three parts in a thousand.

26. Dr Filon's results for torsion.   Dr Filon* considers a cylinder AA' (Fig. 10) of length $2c$ and radius a and supposes that a uniform tangential stress of T dynes per square cm. is applied to the surface over the bands AB and $A'B'$, each of width $\frac{1}{2}c$, in the

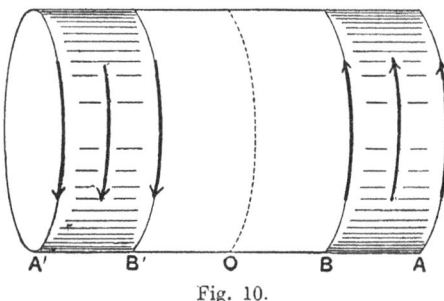

Fig. 10.

manner indicated by the arrows, the total torsional couple being thus equal to $\pi a^2 c T$ dyne-cm.

* "On the elastic equilibrium of circular cylinders under certain practical systems of load." *Phil. Trans. Royal Society*, Vol. 198, A, pp. 147—233.

By symmetry, the particles in the central transverse plane through O, where $z = 0$, suffer no rotation round the axis.

If couples $\pm \pi a^2 c T$, were applied to the plane ends of the cylinder in the manner described in § 39, they would produce uniform torsion and would cause the point A, for which z, the distance of a point from the central plane, is equal to c, and r is equal to a, to move round the circumference of the cylinder through a distance v_0, where

$$v_0 = 2c^2 T/na$$

and n is the rigidity*. Under this uniform torsion, a point, for which $z = pc$ and $r = qa$, would move through a distance pqv_0 at right angles to the plane containing the axis and the radius r. In the actual case, when the torsion is caused by stresses applied to the bands AB and $A'B'$, this particle will move through a distance v at right angles to the plane containing the axis and the radius.

Values of v/v_0.

z	$r = 0$	$r = 0 \cdot 2a$	$r = 0 \cdot 4a$	$r = 0 \cdot 6a$	$r = a$
0	0·0000	0·0000	0·0000	0·0000	0·0000
0·1c	0·0000	0·0197	0·0396	0·0598	0·1003
0·2c	0·0000	0·0392	0·0789	0·1193	0·2007
0·3c	0·0000	0·0584	0·1176	0·1784	0·3018
0·4c	0·0000	0·0767	0·1549	0·2361	0·4046
0·5c	0·0000	0·0937	0·1898	0·2907	0·5169
0·6c	0·0000	0·1086	0·2206	0·3393	0·6190
0·7c	0·0000	0·1210	0·2459	0·3790	0·6920
0·8c	0·0000	0·1301	0·2646	0·4081	0·7430
0·9c	0·0000	0·1360	0·2761	0·4257	0·7734
c	0·0000	0·1376	0·2799	0·4316	0·7835

* The expression for v_0 may be deduced from equation (28), § 89.

Dr Filon has calculated* the values of v/v_0 for the case in which $\pi a = 2c$.

It will be seen from the table that, as long as z or pc is less than $\frac{1}{2}c$, the value of v at $r = qa$ is very nearly equal to pqv_0. Hence between the transverse planes B and B' the particles move round the axis in very nearly the same way as if the cylinder were uniformly twisted by couples applied to the plane ends in the manner described in § 39.

UNIFORM BENDING OF A ROD.

27. Introduction. When a rod is bent, some of the longitudinal filaments are lengthened and some are shortened, and thus it may be expected that the resistance, which the rod offers to bending, will depend upon Young's modulus for the material. In the following sections we shall show how to calculate approximately, in terms of Young's modulus, the system of forces required to bend a rod, when the bending is uniform and small. The exact solution for *finite* bending has not yet been found by any mathematician. A solution, valid for infinitesimal uniform bending, is given in Note XII. An approximate treatment of non-uniform bending is given in Chapter III (EXPERIMENT 10).

We shall consider the case of a straight rod of uniform section and shall suppose that the rod has a plane of symmetry parallel to its length. Then this plane intersects every transverse section in a straight line, which is an axis of symmetry for that section. Bars of rectangular or circular section are examples of such rods.

Let the reader take a steel rule about 0·1 cm. in thickness and two or three cm. in width and let him bend it. He will then observe that the transverse section does not remain a rectangle. The long sides of that section become curved, their centres of curvature and the centres of curvature of the longitudinal filaments of the rule lying on opposite sides of the rule. Rough observations made with the aid of a straight-edge will show that the radius of curvature of a long side of the transverse section is not more than three or four times as great as the radius of any longitudinal filament. Hence, if we take two sections passing through a normal to the face of the steel rule, one transverse and the other

* *Phil. Trans.* Vol. 198, A, p. 229.

longitudinal, the deformation of the transverse section is of the same order of magnitude as the deformation of the longitudinal section, and consequently must be taken into account.

But this curvature of the transverse fibres does not occur to any appreciable extent when the rod takes the form of a thin strip of metal and the radii of curvature of the longitudinal filaments are small compared with a^2/b, where $2a$ is the width of the strip and $2b$ is its thickness. This case therefore requires special investigation, and will be considered in §§ 35 to 37.

28. Strain and stress in a uniformly bent rod. We shall now investigate the system of forces which must be applied to a uniform rod to bend it so that all the longitudinal filaments are bent into circular arcs in planes parallel to a plane of symmetry of the rod. This plane may now be called the *plane of bending.*

In order that the bending may be *uniform* along the length of the rod, it is necessary that the centres of curvature of all the longitudinal filaments should lie on a straight line perpendicular to the plane of bending. This straight line may be called the *axis of bending.* The uniformity of bending also requires that all the particles, which lay in transverse planes before the rod was bent, lie, after the bending, in corresponding planes passing through the axis of bending. These planes therefore cut the filaments at right angles.

We shall examine the strains and the stresses which exist, when the conditions are such that the sides of any longitudinal filament are entirely free from stress. On account of this condition, each filament will be free to contract or expand in a transverse direction when its length is increased or diminished, exactly as if it were isolated from the rest of the rod. We shall show later that the solution obtained in this way is a good approximation to what occurs when a rod is bent by couples applied at its ends, provided that the radii of curvature of the longitudinal filaments are great compared with a^2/b, where $2a$ is the width of the rod, measured in a direction parallel to the axis of bending, and $2b$ is its thickness*.

---

* The method we shall employ is a modification of that used by Thomson (Lord Kelvin) and Tait. (*Natural Philosophy*, Vol. I. Part II. § 711, New Edition.)

Though we have specified that the sides of the longitudinal filaments are free from stress, we do not exclude the possibility that, to produce equilibrium, it may be necessary to apply to each element of volume of the rod a force like that due to gravity, which may be regarded as acting at a distance. Such an action will be called a "body force" and will be measured in dynes per c.c.

Since the rod is uniformly bent, it follows that the stress on a transverse section of any longitudinal filament is normal to that section, and thus the stress is a positive or negative tension.

If the cross-section of any filament be α square cm. and if the stress be a positive tension of T dynes per square cm., the force acting across the section is $T\alpha$ dynes. If the radius of curvature of the filament be r cm. and if the angle between the transverse planes at the ends of a small portion of the filament of length s cm. be θ radians, we have $\theta = s/r$. The resultant of the two forces which are applied to the ends of this portion is a force $2T\alpha \sin \frac{1}{2}\theta$ dynes at right angles to the filament and parallel to the plane of bending. When θ is infinitesimal, the force becomes $T\alpha\theta$ or $T\alpha s/r$, a force proportional to the volume, αs, of the portion of the filament.

To maintain equilibrium a radial "body force" of T/r dynes per c.c. must, therefore, be supplied by some agent acting "at a distance." As such a force does not exist in nature, we shall consider, in §§ 32 to 34, what effects are produced when this force is not supplied. For the present it will be supposed to act.

29. Change of cross-section due to bending. One of the filaments in the plane of bending will be unchanged in length when the rod is bent. This filament will be called the *neutral filament*. Let the radius of curvature of the circular arc into which it is bent be ρ. In Fig. 11* is shown a section of the rod, when bent, made by a plane containing RH, the axis of bending, and here O is the point where the neutral filament cuts this plane. Take rectangular axes OX, OY parallel and perpendicular to RH: let $PN = x$ and $PM = y$ be the coordinates of any point P of the strained section and let PM meet RH in K. Then $MK = \rho$. The

---

* For the sake of clearness, the dimensions of the section have been greatly increased relatively to the distances OR, OS.

longitudinal filament through M is unstretched, since it is at the same distance from RH as the longitudinal filament through O,

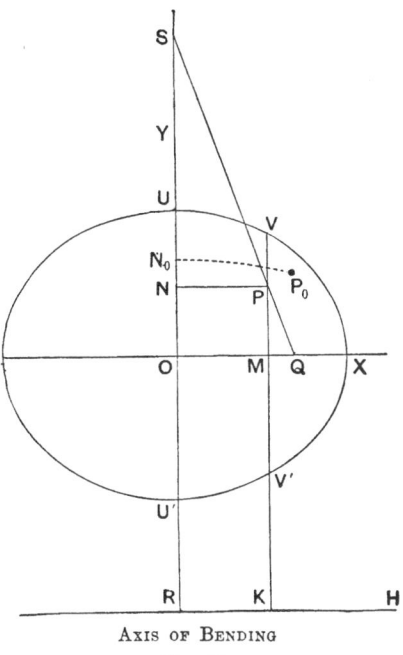

AXIS OF BENDING

Fig. 11.

and hence s', the length, when stretched, of any portion of the filament through P, is to s, its length, when unstretched, as PK is to MK or as $\rho + y$ is to ρ. Hence, if the elongation of the longitudinal filament through P be e, we have

$$e = \frac{s' - s}{s} = \frac{(\rho + y) - \rho}{\rho} = \frac{y}{\rho}. \quad \ldots\ldots\ldots\ldots (1)$$

If f be the lateral contraction of the filament and σ be Poisson's ratio, we have, by § 18, Chapter I,

$$f = \sigma e = \sigma y / \rho. \quad \ldots\ldots\ldots\ldots\ldots\ldots\ldots (2)$$

We can obtain a general idea of how the cross-section is distorted when the rod is bent, if we remember that, since the sides of every longitudinal filament are free from stress, the geometrical form of the cross-section of the *filament* remains unchanged. Thus, if the cross-section be square, it remains square, though changed in area. Hence any two transverse fibres which intersected at any angle in the unstrained section intersect at the same angle in the strained section, and thus the two sets of fibres which were originally parallel to OX and OY are changed by the strain into two sets of curves, such that the curves of one set intersect the curves of the other set at right angles, like lines of force and equipotential lines.

Now those longitudinal filaments, which cut the plane XOY (Fig. 11) above OX when the rod is bent, have suffered a lateral contraction and those below OX have suffered a lateral expansion, and hence the transverse fibres in the plane XOY which were originally parallel to OY are no longer parallel to OY. Further, since the angles of intersection remain unchanged, the transverse fibres which were parallel to OX are bent so that they become curves convex to RH, the axis of bending.

Now, by (2), the lateral contraction of all the longitudinal filaments which are *finally* at the same height above OX is the same, and thus, if the *curved arc* $P_0 N_0$ represent that fibre in the straight rod which is represented by the *straight line* PN in the bent rod, we have, since $PN = x$,

$$\frac{\text{arc } P_0 N_0 - x}{\text{arc } P_0 N_0} = f = \sigma e = \frac{\sigma y}{\rho},$$

or

$$\frac{x}{\text{arc } P_0 N_0} + \frac{\sigma y}{\rho} = 1.$$

Hence, all the points in the section of the straight rod, whose distances from OY, measured along curves of the same character as $P_0 N_0$, were constant and equal to l, now lie upon the straight line

$$\frac{x}{l} + \frac{\sigma y}{\rho} = 1.$$

Any fibre such as $P_0 N_0$ in the straight rod will have a definite radius of curvature q in the neighbourhood of N_0. If we treat $P_0 N_0$ as part of a *circle* of radius q and denote by x_0 the perpendicular distance of P_0 from OY, we have

$$x_0 = q \sin (l/q),$$

or

$$l = q \sin^{-1}\left(\frac{x_0}{q}\right) = q\left(\frac{x_0}{q} + \frac{1}{2}\frac{x_0^3}{3q^3} + \frac{1 \cdot 3}{2 \cdot 4}\frac{x_0^5}{5q^5} + \dots\right).$$

Thus approximately, when x_0/q is small,

$$\frac{l - x_0}{x_0} = \frac{x_0^2}{6q^2}. \quad \dots\dots\dots\dots\dots\dots\dots(3)$$

Hence, if $x_0 = q/10$, the difference between l and x_0 is approximately $x_0/600$ and thus, when x_0 does not exceed $q/10$, we may neglect the difference. In this case we may consider that the transverse fibre in the unbent rod which is straight and has the equation

$$x_0 = \text{constant}$$

is transformed by the bending of the rod into the straight fibre aving the equation

$$\frac{x}{x_0} + \frac{\sigma y}{\rho} = 1. \quad \dots\dots\dots\dots\dots\dots\dots(4)$$

This straight line is represented by SQ in Fig. 11. It cuts OX in Q, where $OQ = x_0$, and OY in S, where $OS = \rho/\sigma$. Thus, to our degree of approximation, OQ is equal to the original length of the fibre which is represented by PN in the bent rod. Since ρ/σ is independent of x_0, the point S is fixed, and hence all the straight lines which are parallel to OY in the unstrained section are changed into straight lines passing through S. If we denote the distance OS by ρ', we have

$$OS = \rho' = \rho/\sigma. \quad \dots\dots\dots\dots\dots\dots\dots(5)$$

Since the lines originally parallel to OX are strained so as to cut at right angles the lines which were originally parallel to OY, it follows that the former are changed by the strain into arcs of circles having S as a common centre.

Hence the transverse fibre passing through O and originally perpendicular to OY is strained into the form of a circle of radius $\rho' = \rho/\sigma$.

When y is small compared with ρ, the radius of curvature q of any transverse fibre in the straight rod (such as $P_0 N_0$), which becomes parallel to OX in the bent rod, is approximately equal to ρ', and hence (3) may be written

$$\frac{l - x_0}{x_0} = \frac{x_0^2}{6\rho'^2}.$$

Thus the above investigation applies with great accuracy so long as ρ' is not less than 10 times the greatest value of x_0.

The longitudinal filaments which lay in the plane through O perpendicular to OY, when the rod was unstrained, are strained so as to lie upon an anticlastic surface having radii of curvature ρ and ρ' in the principal sections at O, the two centres of curvature, R and S, lying on opposite sides of the surface *.

To complete the investigation of the change of form of the transverse section, we will find how the distance η between O and any point N on the axis OY in the strained section depends on the distance y_0 between the corresponding points in the unstrained section. If on OY we take a neighbouring point defined by $\eta + d\eta$, its distance from the first point is changed by the strain from dy_0 to $d\eta$, and hence the lateral contraction f is

$$f = \frac{dy_0 - d\eta}{dy_0}.$$

Hence by (2) and (5)

$$\frac{d\eta}{dy_0} = 1 - f = 1 - \frac{\sigma\eta}{\rho} = 1 - \frac{\eta}{\rho'}. \quad \ldots\ldots\ldots\ldots(6)$$

Now, in the case of metals, Hooke's law begins to fail when the elongation e exceeds about $\frac{1}{1000}$ and thus we see from (1) that, in practical measurements, ρ should be so large in comparison with the thickness of the rod in the plane of bending that the maximum value of η/ρ does not exceed $\frac{1}{1000}$. Thus since, by § 19, Chapter I, σ cannot be greater than $\frac{1}{2}$ in an isotropic elastic solid, $\sigma\eta/\rho$ will not exceed $\frac{1}{2000}$ and thus it will suffice to write y_0 for η on the right side of (6). Then, since $\eta = 0$ when $y_0 = 0$, we obtain, on integration,

$$\eta = y_0 - \frac{y_0^2}{2\rho'}. \quad \ldots\ldots\ldots\ldots\ldots(7)$$

The exact solution is easily found to be

$$\eta = \rho'(1 - e^{-\frac{y_0}{\rho'}}) = y_0 - \frac{y_0^2}{1.2\rho'} + \frac{y_0^3}{1.2.3\rho'^2} - \ldots,$$

which is nearly the same as (7) when y_0/ρ' is small.

We can now construct a diagram to show the distortion of the transverse section of a rectangular rod. In Fig. 12, O is the point

---

* Fig. 18 may assist the reader to realise the character of an anticlastic surface.

where the neutral filament cuts the plane of the paper, and S is the centre of curvature of those *transverse* filaments which are initially parallel to the axis of bending. The sides AB, CD, which were initially parallel to OS, become straight lines $A'B'$, $C'D'$ passing through S, the distance OS being equal to ρ' or ρ/σ, while the sides BC, AD become circular arcs with S as centre.

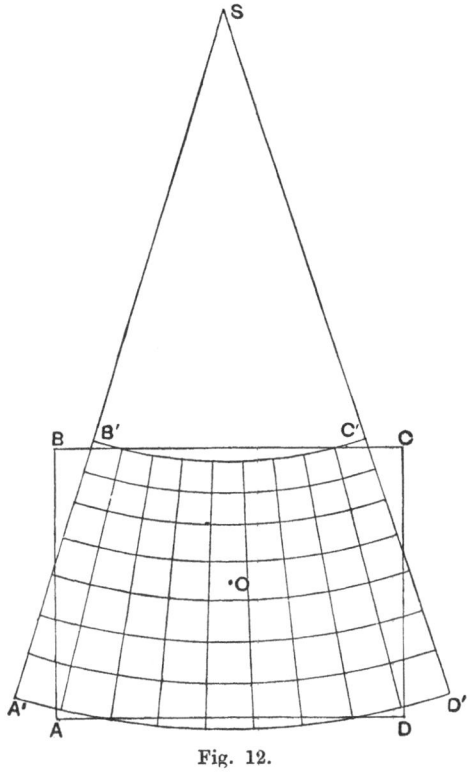

Fig. 12.

Two sets of straight lines parallel to AB and to BC and dividing the section into equal infinitesimal squares is transformed by the strain into a set of radii and a set of circular arcs which divide the strained section into infinitesimal squares. The area enclosed by a mesh of the strained net-work increases as we pass from $B'C'$ to $A'D'$, as appears from (6).

The distortion of the cross-section of a rod of *any* section can be found by the aid of Fig. 12.   On the net-work of straight lines drawn on the section of the unstrained rod we mark a set of points lying on the boundary of the unstrained section, and then mark on the net-work of radii and circular arcs a second set of points corresponding to the first set.   The curve drawn through the second set is the boundary of the strained section.

30.   Position of the neutral filament.   The resultant effect of the normal stresses across any transverse section may be reduced to (a) a force F which acts at right angles to the section and will be taken as acting along the tangent to the neutral filament, and (b) a couple G, having its axis perpendicular to the plane of bending*.

Under the assumed conditions, the sides of a longitudinal filament are free from stress and thus, by (5), § 17, Chapter I, if T be the tension in a longitudinal filament. $T = Ee$.   Hence, by (1)

$$T = Ey/\rho. \quad \dots\dots\dots\dots\dots\dots\dots(8)$$

If α be an element of the strained section at a distance y from OX, we have

$$F = \Sigma T\alpha = \frac{E}{\rho}\Sigma\alpha y. \quad \dots\dots\dots\dots\dots\dots(9)$$

The resultant of the two forces which act at the ends of a portion of the bent rod comprised between two transverse planes inclined at an infinitesimal angle θ is $F\theta$, and hence, since this portion corresponds to a length $\rho\theta$ of the neutral axis, we see that the resultant of the body forces per unit of length of the neutral axis is F/ρ.

Now *any* distribution of normal stress over the transverse section such that $T = Ey/\rho$, together with the body force T/r or $T/(y + \rho)$ per unit volume will bend the rod in such a way that the sides of the longitudinal filaments are free from stress, but to different distributions there will correspond different neutral filaments.

The most important distribution of stress is that for which the force F vanishes, for then F/ρ, the resultant of the body forces

---

* See Note I.

per unit length of the neutral axis, vanishes also. In this case the body forces could be supposed to arise from mutual actions occurring within the rod and would not require the operation of any external agent.

The force F will vanish provided that

$$\Sigma \alpha y = 0. \qquad \qquad (10)$$

But $\Sigma \alpha y = Ah$, where A is the area of the strained section and h is the distance of the "centre of gravity" or centroid of the section above OX. Hence $h = 0$, or, in other words, the neutral filament passes through the centre of gravity of the strained section. When the deformation of the transverse section is very small, we may consider that the longitudinal filament through the centre of gravity of the *unstrained* section remains unchanged in length and is therefore the neutral filament.

Since the force F vanishes, the stresses acting across any transverse section are equivalent to a couple.

31. Bending moment. By § 4, Chapter I, the sum of the moments about the axis OX (Fig. 11) of the forces exerted on the part of the rod on *one* side of the transverse section by the tensions in the longitudinal filaments is equal and opposite to the "bending moment," i.e. the moment about the same axis of the forces applied to the same part of the rod. The force due to the tension T in a filament of section α is $T\alpha$ and this force acts at a distance y from OX. Hence, by (8), if G be the bending moment,

$$G = \Sigma T\alpha y = \frac{E}{\rho} \Sigma \alpha y^2.$$

But $\Sigma \alpha y^2$ is the "moment of inertia" of the area of the strained section about the axis OX. If $\Sigma \alpha y^2$ be denoted by I, we have*

$$G = \frac{EI}{\rho}. \qquad \qquad (11)$$

When the rod is bent by *couples* applied to its ends, the neutral filament passes through the centre of gravity of the strained section (§ 30) and then I is the "moment of inertia" of the strained section about an axis through its centre of gravity perpendicular to the plane of bending.

---

* The values of I for some simple forms of area are given in Note IV, § 12.

Since the deformation of the transverse section is very small in practical work, we may take I as equal to I_0 the "moment of inertia" of the unstrained section about an axis through its centre of gravity at right angles to the plane of bending, unless the rod takes the form of a blade. In this case (11) no longer gives the couple when $1/\rho$ becomes at all large compared with b/a^2, where $2a$ is the width of the blade parallel to OX and $2b$ is its thickness. We shall see in § 37 that we must now write

$$G(1 - \sigma^2) = EI_0/\rho.$$

Equation (11) is of fundamental importance in the theory of the bending of rods and is frequently required in practical work.

32. Removal of the "body-forces." We must now examine the effects which would follow the removal of the "body forces" which were introduced in § 28 to ensure that the sides of the longitudinal filaments should be free from stress. The removal may be effected by superposing a second set of body forces equal in magnitude to those already applied, but with opposite directions. We may regard the results of §§ 29, 30 and 31 as good approximations to the results corresponding to the natural case in which a rod is bent by couples applied to its ends and is not acted on by any body forces, provided that the form found for the strained section in § 29 is not perceptibly changed when the second set of body forces is applied.

In Fig. 11, the coordinates of P are x, y and hence, by § 28, if Y be the body force at P,

$$Y = T/(\rho + y) \text{ dynes per c.c.}$$

But, by (8), $T = Ey/\rho$

and hence $$Y = \frac{Ey}{\rho(\rho + y)}.$$

Now the volume of a portion of a longitudinal filament which has the cross-section $dx\,dy$ and is terminated by a pair of transverse planes inclined at a small angle θ is $(\rho + y)\theta\,dx\,dy$, and hence the force acting on this element is

$$Y(\rho + y)\theta\,dx\,dy \text{ dynes,}$$

or $$E\theta y\,dx\,dy/\rho. \quad\ldots\ldots\ldots\ldots\ldots\ldots\ldots\ldots(12)$$

Thus the force is proportional to y and acts upwards or downwards according as P is above or below OX. When the rod is bent by couples, the resultant of the "body forces" on any portion bounded by a pair of transverse planes is zero, and hence, since OY is a line of symmetry for this section, the "body forces" which act on the part of the wedge of angle θ which lies on one side of OY must themselves have a zero resultant and are thus equivalent to a couple.

If we consider the total force acting on an elementary wedge of angle θ and width dx, extending from V to V' (Fig. 11), we see that it vanishes when $MV = MV'$ and that it will nearly vanish when MV and MV' are nearly equal.

When the cross-section is initially symmetrical with respect to OX, we see, by § 29, that in the neighbourhood of OY, MV is less than MV' and hence, since O is the centre of gravity of the strained section, MV is greater than MV' when PN exceeds some definite value*. Hence the *reversed* forces which correspond to the part of the section to the right of OY will give rise to a couple tending to destroy the curvature of the transverse fibres which had been bent into circular arcs, and will cause new stresses in these transverse fibres. It would be difficult to determine the precise values of these new stresses, but it is clear that there will be a positive tension in the transverse fibres near U (Fig. 11) and a negative tension in the transverse fibres near U'.

To make the discussion as definite as possible we shall consider the case of a bar of rectangular section, having a width $2a$ parallel to OX, i.e. parallel to the axis of bending, and a thickness $2b$ parallel to OY.

33. Case of a rod. If a straight line be drawn between the ends of the circular arc passing through the point O (Fig. 12), the greatest distance of the arc from the chord is approximately $a^2/2\rho'$; when this is small MV and MV' (Fig. 11) will be nearly equal, since O is the centre of gravity of the strained section. In this case the bending moment due to the reversed "body forces" will be small. If, at the same time, b be large compared with $a^2/2\rho'$, we may expect that the change of section due to the reversed

---

* These statements are illustrated in Fig. 12.

"body forces" will be small, in which case the distortion represented by Fig. 12 will be a good approximation to that which actually occurs when a rod is bent by couples and no "body force" is supplied.

In the Practical Example of EXPERIMENT 9, Chapter III, the rod was such that $a = 1·24$, $b = 0·15$ cm. and thus $a^2/2b = 5·13$ cm. The least value of ρ' in the experiments was 485 cm., or nearly 100 times $a^2/2b$. In such a case the effect of the reversed "body forces" may be safely neglected.

34. Case of a blade. When a/b is great, a quite moderate bending of the rod is sufficient to cause ρ' to be small compared with $a^2/2b$, although ρ' may be great compared with a as well as with b. Thus, it is quite easy to bend a blade or strip of thin metal, for which $b = 0·01$ cm., so that ρ, the radius of curvature of the longitudinal fibres, is 10 cm. The value of ρ' deduced from the formula $\rho' = \rho/\sigma$ is then about 30 or 40 cm. according to the value of σ. But a need be no greater than 2 cm. to make $a^2/2b = 200$ cm., and then $a^2/2b$ much exceeds ρ', and $a^2/2\rho'$ much exceeds b.

The case in which ρ is small compared with $a^2/2b$ is of some interest, since, under these conditions, the actual distortion of the

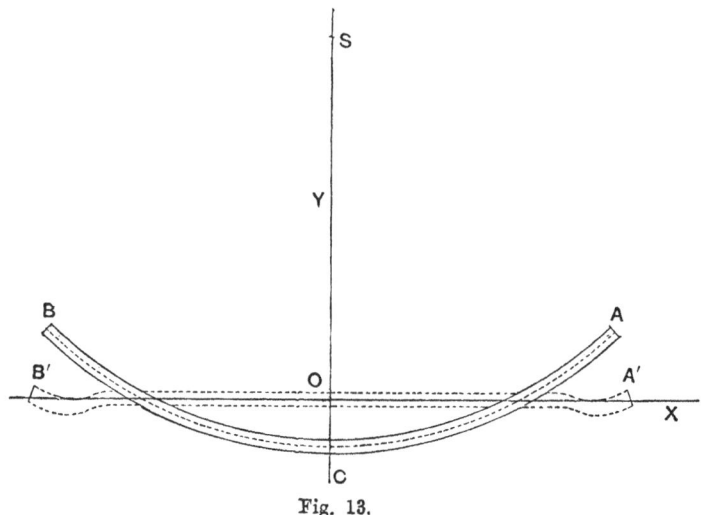

Fig. 13.

cross-section, when the blade is bent, differs entirely from that discussed in § 29 and represented in Fig. 12. We shall therefore consider this case in some detail.

Let AB (Fig. 13) represent the section of the bent blade and let C be the point on OY which is midway between the curved sides of the section. By § 29, we see that, when a/ρ' is small, we may take the dotted curve ACB, which is midway between the sides of the section, to be an arc of a circle of radius ρ'.

We must first find the distance of C from O, the centre of gravity of the strained section. If $OC = p$, the curve ACB may be represented approximately by

$$y = \frac{x^2}{2\rho'} - p.$$

To find p we express the fact that the centre of gravity coincides with O. Thus approximately, if h be the ordinate of the centre of gravity of the section ACB,

$$4abh = 2\int_0^a y \cdot 2b \, dx = 4b \int_0^a (x^2/2\rho' - p) \, dx$$

$$= 4ab \left(\frac{a^2}{6\rho'} - p\right).$$

But $h = 0$, and hence

$$p = a^2/6\rho',$$

so that the equation to ACB is

$$y = (\tfrac{1}{2}x^2 - \tfrac{1}{6}a^2)/\rho'.$$

Let M be the moment about an axis through C, perpendicular to the plane of Fig. 13, of the "body forces" which act on the half (corresponding to AC) of a portion of the blade bounded by two transverse planes inclined at a small angle θ. Then if we write $2b$ for dy in (12), we find

$$M = \frac{E\theta}{\rho} \int_0^a xy \cdot 2b \, dx = \frac{2bE\theta}{\rho\rho'} \int_0^a (\tfrac{1}{2}x^3 - \tfrac{1}{6}a^2 x) \, dx$$

$$= \frac{a^4 b E\theta}{12\rho\rho'}.$$

The section of this portion of the blade in a plane through OY perpendicular to the plane of Fig. 13 is approximately a rectangle with sides $\rho\theta$ and $2b$, and thus, if J be the moment of inertia of this rectangle about an axis through C perpendicular to the plane of the figure*,

$$J = \tfrac{2}{3} b^3 \rho\theta.$$

In finding $1/R$, the change of transverse curvature which the *reversed* "body forces" would produce in the portion of the blade under consideration, we must remember that the fibres at right angles to the plane of Fig. 13 are unable to change in length owing to their connexion with the neighbouring parts of the blade, and we must make allowance for this in the manner described in §§ 35, 36, 37. We can then make a rough estimate of R by applying the formula (22) of § 37 to this case and writing

$$\frac{EJ}{(1 - \sigma^2) R} = M.$$

Hence

$$R = \frac{EJ}{(1 - \sigma^2) M} = \frac{8\rho^2 b^2}{(1 - \sigma^2) a^4} \cdot \rho'.$$

If $\sigma = \tfrac{1}{4}$, R will equal ρ' when $\rho^2 = 15 a^4 / 128 b^2$ or $\rho = 0\cdot34 a^2 / b$. Hence we may expect that, when ρ is less than $a^2/3b$, the section of the blade, instead of being bounded by radii and arcs of concentric circles will be practically a rectangle $A'OB'$ with a slight distortion at each end as shown by the dotted lines in Fig. 13. The reader may easily confirm this expectation by bending a thin strip of metal so that ρ is less than $a^2/3b$. It will be seen that the anticlastic curvature of Fig. 12 no longer exists.

UNIFORM BENDING OF A BLADE.

35. Introduction. In §§ 28 to 33 we have investigated the bending of a rod when the sides of the longitudinal filaments are free from stress and have found that the cross-section will be distorted as in Fig. 12. But experiment shows that when a blade, i.e. a long and wide strip of thin metal, is uniformly bent, it does not differ appreciably from part of a circular cylinder, however great the curvature may be, and thus the transverse fibres originally parallel

* See Note IV, § 12.

to the axis of bending are not appreciably bent. Since, for the reasons given in § 34, the theory of § 29 entirely fails in this case, a fresh investigation is required.

We shall now consider the bending of a blade when the conditions are such that the transverse fibres originally parallel and perpendicular to the axis of bending remain parallel and perpendicular to that axis after bending. As in § 28, we shall introduce a "body force" to counterbalance the radial force due to the tensions on the ends of any element of a longitudinal filament.

The blade before it is bent is a rectangular block of length $2l_0$, of width $2a_0$ and of thickness $2b_0$, and a_0 is great compared with b_0. When it is bent, the filaments parallel to the length $2l_0$ lie along circular arcs.

Let $ABCD$ (Fig. 14) be a section of the blade when bent, the side AD being parallel to RH, the axis of bending. Let $AD = 2a$ and $AB = 2b$. Since all the transverse fibres perpendicular to RH remain straight and perpendicular to RH, the lateral expansion of every part of each of these fibres parallel to RH is the same. This expansion will be denoted by u cm. per cm.

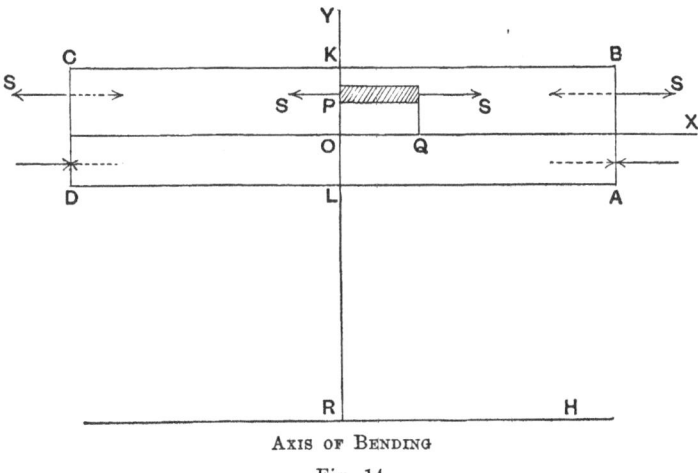

AXIS OF BENDING

Fig. 14.

Let OX be the straight line intersected by all the unstretched *longitudinal* filaments and let the small rectangle P represent an element of area of unit length OQ and width dy, where $PO = y$

Since the element P retains its rectangular form, the stresses on the sides of the longitudinal filament of which P is a section must be normal to those sides. The introduction of the "body force" relieves the sides parallel to OX of stress and there only remains a normal stress on the sides parallel to OY. This we shall suppose is a tension of S dynes per square cm. The longitudinal tension of the filament is T dynes per square cm.

Just as in § 29, if the elongation of the longitudinal filament through P be e,

$$e = y/\rho,$$

where $OR = \rho$. This elongation is due to the stresses T and S alone, since there is no pressure or tension in a direction at right angles to both T and S. The stress T causes an elongation T/E and the stress S a contraction $\sigma S/E$, both in the direction of T. Thus

$$T - \sigma S = Ee = Ey/\rho. \quad \ldots\ldots\ldots\ldots\ldots(13)$$

Similarly, the tension S causes an elongation S/E and the tension T a contraction $\sigma T/E$ parallel to OX, and thus

$$-\sigma T + S = Eu. \quad \ldots\ldots\ldots\ldots\ldots(14)$$

From (13) and (14) we obtain

$$(1 - \sigma^2)\, T = Ey/\rho + \sigma Eu, \ldots\ldots\ldots\ldots(15)$$

$$(1 - \sigma^2)\, S = \sigma Ey/\rho + Eu. \ldots\ldots\ldots\ldots(16)$$

If we could assume that the width of the blade remains unchanged, so that $u = 0$, we could at once find T and S in terms of y. But there do not appear to be any grounds for this assumption and hence the value of u must be found. The calculation shows, however, that u is negligible, being of the *second* order of small quantities.

36. Position of neutral filaments. If h be the height of the centre of gravity of the area $ABCD$ above OX, and if BC and AD cut OY in K and L respectively, we have $OK = b + h$ and $OL = b - h$, and thus

$$\int_{h-b}^{h+b} dy = 2b, \qquad \int_{h-b}^{h+b} y\, dy = 2hb.$$

Further, if I be the moment of inertia of the section about OX,

$$I = 2a \int_{h-b}^{h+b} y^2 dy = 4ab\,(h^2 + \tfrac{1}{3}b^2).$$

When the blade is bent by couples, the resultant of the stress T over the area $ABCD$ is zero. Using the first two of the above integrals, we find by (15) that

$$(1 - \sigma^2) \int_{h-b}^{h+b} T \cdot 2a\,dy = \left\{ \frac{2Ehb}{\rho} + 2\sigma Ebu \right\} 2a.$$

But the integral on the left side is zero, since the resultant force is zero, and hence

$$u = -\,h/\sigma\rho. \quad\quad\dots\dots\dots\dots\dots\dots(17)$$

When the blade is bent by couples, the total force across a section of the blade made by a plane perpendicular to the axis of bending is zero, and thus, since the area of the curved strip of this section corresponding to dy is proportional to $\rho + y$, we have

$$\int_{h-b}^{h+b} S\,(\rho + y)\,dy = 0. \quad\quad\dots\dots\dots\dots(18)$$

Multiplying (16) by $\rho + y$ and integrating, we find

$$(1 - \sigma^2) \int_{h-b}^{h+b} S\,(\rho + y)\,dy = \frac{2b\sigma E}{\rho}\,(h\rho + h^2 + \tfrac{1}{3}b^2) + 2bEu\,(\rho + h) = 0,$$

so that

$$u = -\,\frac{\sigma\,(h\rho + h^2 + \tfrac{1}{3}b^2)}{\rho\,(\rho + h)}\,. \quad\quad\dots\dots\dots\dots(19)$$

Substituting the value of u given by (17), we obtain a quadratic equation for h. Thus

$$h^2 + h\rho - \frac{\sigma^2 b^2}{3\,(1 - \sigma^2)} = 0.$$

Hence

$$h = \frac{\rho}{2} \left[-1 \pm \left\{ 1 + \frac{4\sigma^2 b^2}{3\,(1 - \sigma^2)\,\rho^2} \right\}^{\frac{1}{2}} \right].$$

Since h is very small in comparison with ρ, we select the positive sign for the square root. Now the second term under the square root is small compared with unity, and thus, by expanding, we obtain the approximate value

$$h = \frac{\sigma^2 b^2}{3\,(1 - \sigma^2)\,\rho}\,. \quad\quad\dots\dots\dots\dots\dots(20)$$

When the bending is slight, h is very small. Thus, if $\sigma = \frac{1}{4}$, we have

$$h/b = b/45\rho.$$

By (17), the lateral expansion, u, is given by

$$u = -\frac{h}{\sigma\rho} = -\frac{\sigma b^2}{3\,(1 - \sigma^2)\,\rho^2}. \quad\ldots\ldots\ldots\ldots(21)$$

When the bending is so slight that b/ρ is a small quantity of the first order, u is a small quantity of the second order and may be neglected. Hence, to this order of accuracy, we may say that the width of the blade parallel to the axis of bending remains unchanged.

37. Bending moment. If the bending moment about the axis OX be G dyne-cm., we have, since, by (17), $\sigma u = -h/\rho$,

$$G = 2a \int_{h-b}^{h+b} Ty\,dy = \frac{2a}{1-\sigma^2} \int_{h-b}^{h+b} \left(\frac{Ey}{\rho} + \sigma Eu\right) y\,dy$$

$$= \frac{2aE}{1-\sigma^2} \left\{\frac{2b}{\rho}\,(h^2 + \tfrac{1}{3}b^2) - \frac{h}{\rho}\,2hb\right\}$$

$$= \frac{4ab^3 E}{3\,(1-\sigma^2)\,\rho}.$$

This is the *accurate* value of G in terms of $2a$ and $2b$, the sides of the section of the bent blade. When ρ is great compared with b, we may replace a and b by a_0 and b_0, the values for the unbent blade, and may neglect h^2 in comparison with $\frac{1}{3}b^2$. Then I takes the value $I_0 = \frac{4}{3}a_0 b_0^3$. Hence, when b_0/ρ is small,

$$G = \frac{EI_0}{(1 - \sigma^2)\,\rho}. \quad\ldots\ldots\ldots\ldots\ldots(22)$$

Thus G is greater than the value EI_0/ρ given by the theory of § 31 in the proportion of 1 to $1 - \sigma^2$. The difference is, however, never great, since σ is about $\frac{1}{4}$ for metals and cannot exceed $\frac{1}{2}$ in isotropic solids.

When the bending is so slight that b/ρ is small, we may neglect the effects of the "body force," since this is equal to $T/(\rho + y)$ or approximately to $Ey\,(1 - \sigma^2)^{-1}\,\rho^{-2}$, and can be made as small as we please compared with T and S by sufficiently increasing ρ.

But when we bend a blade by couples applied to its ends, the tensions S indicated by the arrows (Fig. 14) are not applied to the edges AB and CD. We can however correct our solution by applying to AB and CD a set of tensions equal to S but with their directions reversed, as shown by the dotted arrows. These reversed tensions will tend to change the section of the blade, but as soon as the section is changed the changed tension in the longitudinal filaments near the edge of the blade will give rise to radial forces tending to counteract the effects of the reversed tensions. Experiment shows that any distortion which the section may suffer near the edges AB and CD is exceedingly small and that it cannot be made appreciable by increasing the curvature of the longitudinal filaments. We therefore conclude, that, when the width of the blade is great compared with its thickness, a pair of couples applied to the ends of the blade will bend it so that its surfaces do not differ appreciably from the cylindrical form, provided that the radius of curvature be small enough to make the product of the radius and the thickness of the blade small compared with the square of the width of the blade. In this case we may regard (22) as giving a nearly accurate value for the bending moment.

38. Change of type of bending. When the bending is very slight, so that a^2/ρ is very small compared with b, the section of the blade will be changed in the manner described in § 29. The bending moment will then be connected with the curvature by the equation

$$G = \frac{EI_0}{\rho}.$$

But, when the bending is increased so that a^2/ρ becomes large compared with b, there will be no appreciable change of section, and the bending moment will now be

$$G = \frac{EI_0}{(1 - \sigma^2)\,\rho}.$$

The blade is consequently a little less stiff for small curvatures than for large ones. As the curvature is increased from zero to a large value, the product $G\rho$ will gradually change from EI_0 to $EI_0/(1 - \sigma^2)$.

It would be interesting to endeavour to detect experimentally the change in the product $G\rho$ as the bending proceeds. If the method of EXPERIMENT 12, Chapter III, be employed, success will largely depend upon a proper choice of the section of the blade and upon the use of a *sensitive* and accurate instrument for measuring the displacement of the centre of the blade. The bending of the blade under its own weight should be made as small as possible by the use of a short blade and by a proper choice of the distance between the knife edges.

The blade used in the Practical Example of EXPERIMENT 12 is not suitable for the suggested experiment. For this blade,

$$a = 2\cdot521 \text{ cm.}, \quad b = 0\cdot02406 \text{ cm.}$$

and hence $a/b = 105$ and $a^2/b = 264$ cm.;

thus ρ must exceed 2000 cm. if it is to be considered great compared with a^2/b. The unavoidable curvature due to the weight of the blade is of the same order of magnitude as $1/2000$ cm.$^{-1}$, and thus it will be understood that an attempt to detect the change of type of bending led to no result. A blade with a larger value of a/b should be used.

UNIFORM TORSION OF A ROUND ROD.

39. Relation between torsional couple and twist. Consider a round rod or wire of length l cm. and radius a cm., having plane ends A, B, at right angles to the axis, and let us enquire if it be possible to apply such a distribution of forces to the rod that it shall suffer a uniform torsion, in which the distances of every particle from the axis and from the plane A remain unchanged and all the particles in any one normal section describe equal angles about the axis. If the particles at A be fixed and those at B describe angles of ϕ radians about the axis, the twist per unit length is ϕ/l radians per cm.

Consider a portion AC (Fig. 15) of the rod in the form of a thin tube of length h cm. and radii r and $r + dr$ cm., the end A being fixed. A point Q on the end C describes the angle QOQ' or $h\phi/l$ radians about the axis and therefore moves through $hr\phi/l$ cm. relative to the fixed end. Thus the thin prism PQ, which is cut out by a pair of radial planes, is strained into the figure PQ'. Since for strains within the elastic limit the angle QPQ' is very small, we may treat its tangent and its circular measure as identical. Thus $QPQ' = QQ'/QP = r\phi/l$ radians. In other words, the prism has suffered a shear $r\phi/l$ in the plane QPQ'. There is no change

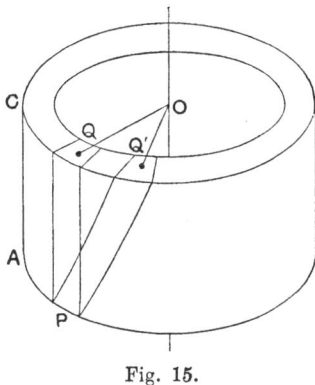

Fig. 15.

of volume, for the radial width, the width measured round the circumference of the cylinder and the height all remain unchanged. There is no shear in a plane containing the axis and no shear in a plane normal to the axis. Hence the shear $r\phi/l$ in the plane QPQ' is the whole strain.

By Chapter I, § 14, this strain can be produced in the prism PQ by tangential stresses $nr\phi/l$ dynes cm.$^{-2}$ acting on the ends P, Q parallel to the plane QPQ', together with tangential stresses of equal amounts parallel to the axis acting on the radial faces. The latter stresses are provided by the action of the neighbouring prisms in the tube. Thus, the only stress on the ends of the prism is the tangential stress $nr\phi/l$ dynes cm.$^{-2}$. If the cross section of the prism be α square cm., the tangential force is $\alpha nr\phi/l$ dynes, and the moment of this about the axis is $\alpha r^2 . n\phi/l$ dyne-cm. If the total moment of all the forces which act across the whole section of the rod be G, we have

$$G = (n\phi/l)\, \Sigma \alpha r^2 \text{ dyne-cm.}$$

It is important to notice that the given strain does not imply the action of any forces on the *cylindrical* surface of the rod.

The quantity $\Sigma a r^2$ is the "moment of inertia" of the area of cross-section of the rod about the axis. Its value can be found at once by the integral calculus. For, if we take $2\pi r\,dr$ as the element of area, we have, for a circle of radius a cm.,

$$\Sigma a r^2 = \int_0^a r^2 . 2\pi r\,dr = \tfrac{1}{2}\pi a^4 \text{ cm.}^4$$

A method of calculating the value of $\Sigma a r^2$ without the use of the calculus is given in Note IV, § 12. Hence

$$G = \frac{\pi n a^4 \phi}{2l} \text{ dyne-cm.} \qquad \ldots\ldots\ldots\ldots\ldots(23)$$

Thus, if we can find by experiment the couple corresponding to the twist ϕ, the value of n can be deduced from (23).

The quantity $\Sigma a r^2$ is sometimes called the "second moment" of the area about the axis.

Since there is no stress on the cylindrical surfaces of an elementary tube, such as that shown in Fig. 15, it follows that the investigation applies to any tube bounded by two circular and coaxal cylinders. If a, b be the radii of the cylinders, we have

$$G = \frac{\pi n\,(a^4 - b^4)\,\phi}{2l} \text{ dyne-cm.}$$

In this calculation it has been assumed that the material is homogeneous and isotropic, a condition improbable in the case of a wire, where the material has been made to "flow" in the wire-drawing process. When the material is not isotropic and homogeneous, there is no such thing as *the* rigidity of the material and, hence, the application of (23) to the experimental value of the ratio of the couple to the twist only leads to a sort of average value of the rigidity, such that an isotropic and homogeneous rod of length l and radius a, formed of material with this rigidity, would offer the same resistance to torsion as the rod used in the experiment.

40. Rods of non-circular section. We have seen that, for a rod of circular section, the couple G and the twist ϕ are connected by the equation

$$\frac{Gl}{n\phi} = \Sigma a r^2.$$

Here Σar^2 is the "moment of inertia" of the section of the rod about the axis of the rod. But it does not follow and it is not true that, for a rod of any other section, $Gl/n\phi$ is equal to the "moment of inertia" of the section. It can be shown that, in the general case, the surfaces, which are initially perpendicular to the axis of the rod, cease to be plane when the rod is twisted, and thus the investigation of § 39 does not apply to the general case. Complete solutions have been obtained for several forms of section, and the following values of $Gl/n\phi$ have been found. For comparison we give in each case the value of I, the "moment of inertia" of the area of the section about an axis through its centre of gravity and perpendicular to its plane (see Note IV, § 12).

It will be seen that in every case, except that of a circular section, $Gl/n\phi$ is less than I, the "moment of inertia" of the section.

Circular area, radius a.

$$\frac{Gl}{n\phi} = \tfrac{1}{2}\pi a^4, \qquad I = \tfrac{1}{2}\pi a^4.$$

Elliptical area, axes 2a, 2b.

$$\frac{Gl}{n\phi} = \frac{\pi a^3 b^3}{a^2 + b^2}, \qquad I = \tfrac{1}{4}\pi ab\,(a^2 + b^2). \quad\ldots\ldots(24)$$

Rectangular area, sides 2a, 2b.

$$\frac{Gl}{n\phi} = \frac{16ab^3}{3} - b^4\left(\frac{4}{\pi}\right)^5 \left\{ \sum_{m=0}^{m=\infty} \frac{1}{(2m+1)^5} \tanh\frac{(2m+1)\,\pi a}{2b} \right\}, \quad\ldots(25)$$

where m has the values $0, 1, 2, 3 \ldots$,

$$I = \tfrac{4}{3}ab\,(a^2 + b^2).$$

For a square, this gives

$$Gl/n\phi = 2\cdot2492a^4, \qquad I = \tfrac{8}{3}a^4 = 2\cdot6666\ldots a^4. \quad\ldots\ldots(26)$$

When a is greater than $3b$, the sum of the infinite series of hyperbolic tangents, which is contained within the brackets in (25), differs by less than two parts in 10,000 from

$$\frac{1}{1^5} + \frac{1}{3^5} + \frac{1}{5^5} + \ldots = 1\cdot00452^*\ldots,$$

its value when a/b is infinite. Thus, when $a > 3b$, we may put

$$Gl/n\phi = ab^3\,(\tfrac{16}{3} - 3\cdot361b/a). \quad\ldots\ldots\ldots\ldots(27)$$

_____

* See Dale, *Five-Figure Tables*, p. 92.

41. Practical approximation. In practical cases it is impossible to apply to the plane ends of the rod the ideal distribution of tangential force, in which the force per unit area at each point is proportional to the distance of the point from the axis. But it is easily understood that any distribution of force over the *cylindrical surface* of the rod near one end will produce at some distance from that end the same strain as the ideal distribution provided that the couples due to the two are equal. Thus, suppose that the ends of the rod AB (Fig. 16) are soldered into two stout blocks, C, D. Then, if the rod be twisted by means of these blocks, forces are applied over the curved surface of the rod between the planes A and E, and the strain at any point P between A and B will not be *quite* the same as if the ideal distribution had been applied to the section A, even though the two distributions have

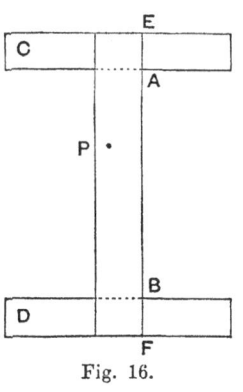

Fig. 16.

the same moment G. But, when AP exceeds two or three diameters, the difference between the strains at P will be inappreciable. For suppose that a couple G is applied to the cylindrical surface of AE by means of the block C and that simultaneously a couple $-G$ is applied to the section A, the force being distributed in the ideal manner. Since these couples are in equilibrium, no couple is required to hold the block D at rest, and, without calculation, we may infer that the strain due to the two opposing couples will be insensible when the distance AP exceeds a few diameters. In other words, the strain at P, due to a couple G applied by the block C, is practically identical with that produced by an equal couple applied in the ideal manner over the section A, provided that PA exceed a few diameters. This result furnishes an illustration of Saint-Venant's principle (§ 24). In practical cases the length of the rod is very many times its diameter and hence, in these cases, it is sufficiently accurate to assume that the uniform torsion extends up to the sections A, B.

Uniform Torsion of a Blade.

42. Introduction. In treatises on the mathematical theory of elasticity, the couple required to produce a given twist in a rod of rectangular section is deduced from the general equations of elasticity by the aid of Fourier analysis, the result being expressed in the form of an infinite series, as in equation (25) of § 40. The use of the Fourier mathematics is unavoidable unless one side of the section is very small compared with an adjacent side. In this case the couple can be calculated by simple methods.

We shall consider a blade of length l, of width $2a$ and of very small thickness $2b$ cm. and shall find the couple (G dyne-cm.) required to twist one end of the blade through an angle of ϕ radians relative to the other end. The twist per centimetre will be denoted by τ; thus

$$\tau = \phi/l \text{ radians per cm.} \quad\ldots\ldots\ldots\ldots\ldots\ldots(28)$$

43. Geometry of a helicoid. We shall first consider the uniform torsion of a strip of a mathematical plane. Let $ABCDA'D'B'$ (Fig. 17) be a rectangular portion of a plane, and let rectangular axes OX, OY, OZ be drawn through the centre O, the axes OX, OZ being perpendicular to the edges of the strip, while the axis OY is perpendicular to the paper and is directed *away from* the reader. Let the strip be now deformed in such a way that a line on the strip initially parallel to OA and at a distance z from OA is turned about the axis of z through an angle τz, the positive direction of rotation being connected with the direction OZ in the same way as the rotation and translation of a right-handed screw working in a fixed nut.

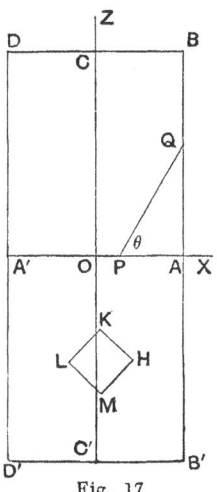

Fig. 17.

The edges of the strip which were initially parallel to OZ thus become uniform right-handed helices making one turn about the axis in a length of $2\pi/\tau$ cm. The new surface is called a helicoid.

If the initial coordinates of a point on the strip are ξ, 0, ζ, they will be changed by the twisting to x, y, z, where

$$x = \xi \cos (\tau \zeta), \quad y = \xi \sin (\tau \zeta), \quad z = \zeta. \quad \ldots\ldots\ldots(29)$$

On eliminating ξ and ζ from these equations, we obtain for the equation to the helicoid

$$y = x \tan (\tau z). \quad \ldots\ldots\ldots\ldots\ldots(30)$$

When τz is very small compared with a radian, we may write τz instead of $\tan (\tau z)$, and thus, in the neighbourhood of OA, the surface may be represented by the equation

$$y = \tau x z. \quad \ldots\ldots\ldots\ldots\ldots\ldots(31)$$

We shall now find the curvature of the helicoid at any point P on OX, where $OP = p$. If we move the origin to P by writing $x' + p$ for x in (31), the equation to the surface becomes

$$y = \tau (x' + p) z. \quad \ldots\ldots\ldots\ldots\ldots(32)$$

Now take a plane containing the new axis of y and cutting the plane OXZ in a straight line PQ inclined at an angle θ to OX. Then, if x', y, z, be the coordinates of a point on the curve of intersection of the plane and the helicoid at a distance r from the new axis of y, we have

$$x' = r \cos \theta, \qquad z = r \sin \theta,$$

and thus, by (32),

$$y = \tau (r \cos \theta + p) r \sin \theta.$$

For a given value of θ, this equation shows the form of the curve of intersection. By § 79, Chapter III, the curvature, $1/\rho$, of this curve at the point P is given by

$$\frac{1}{\rho} = \frac{d^2 y}{dr^2} \left\{ 1 + \left(\frac{dy}{dr}\right)^2 \right\}^{-\frac{3}{2}}.$$

Since we desire the curvature at P, we must put $r = 0$ in this result *after* the differentiations have been performed. We thus obtain

$$1/\rho = 2\tau \sin \theta \cos \theta \{1 + \tau^2 p^2 \sin^2 \theta\}^{-\frac{3}{2}}. \quad \ldots\ldots(33)$$

If τp be so small that $\tau^2 p^2$ is negligible in comparison with unity, the curvature is independent of the position of P and has the value

$$\frac{1}{\rho} = \tau \sin 2\theta. \quad \ldots\ldots\ldots\ldots\ldots(34)$$

It will be seen* that the convexity of the curve of intersection is turned towards the reader when θ lies between 0 and $\frac{1}{2}\pi$ or between π and $\frac{3}{2}\pi$. On the other hand the convexity is turned away from the reader when θ lies between $\frac{1}{2}\pi$ and π or between $\frac{3}{2}\pi$ and 2π.

For the physical applications we require the curvature of the section of the helicoid made by a plane *which contains the normal to the helicoid at P* and is inclined at an angle θ to OX. But, when τp is very small compared with unity, the angle between this normal and the new axis of y is very small, and then (34) will give an approximate value for the curvature of the section of the helicoid made by a plane containing the normal.

By the methods of Solid Geometry we can show, from the exact equation (30), that the accurate expression for the curvature is

$$\frac{1}{\rho} = \frac{\tau \sin 2\theta}{1 + \tau^2 p^2},$$

which agrees with (34) when τp is infinitesimal.

From (34) we see that the curvature vanishes when $\theta = 0$ and when $\theta = \frac{1}{2}\pi$ and that it has the extreme values $\pm \tau$ when $\theta = \pm \frac{1}{4}\pi$. Thus, straight lines initially parallel to OA or OC remain straight, while straight lines initially inclined at $\frac{1}{4}\pi$ to OA are bent to the radius $1/\tau$, the convexity being towards the reader, and those initially inclined at $-\frac{1}{4}\pi$ are bent to the same radius but with their convexity turned away from the reader. Thus the helicoid has anticlastic curvature, and at every point the principal radii of curvature have the constant values $\pm 1/\tau$.

If we take any two neighbouring points on the rectangular strip, the distance between them remains unchanged when the strip is deformed into a helicoid, provided that the distance of either point from the axis be small compared with $1/\tau$. For, if the two points $(\xi, 0, \zeta)$, $(\xi + d\xi, 0, \zeta + d\zeta)$ move to the positions (x, y, z), $(x + dx, y + dy, z + dz)$, we have, by (29),

$$dx = \cos(\tau\zeta)\, d\xi - \tau\xi \sin(\tau\zeta)\, d\zeta$$

$$dy = \sin(\tau\zeta)\, d\xi + \tau\xi \cos(\tau\zeta)\, d\zeta$$

$$dz = d\zeta.$$

* See p. 70.

Hence, if $d\sigma$ and ds be the initial and final distances between the points,

$$(d\sigma)^2 = (d\xi)^2 + (d\zeta)^2,$$

$$(ds)^2 = (d\xi)^2 + (d\zeta)^2 + \tau^2\xi^2(d\zeta)^2 = (d\sigma)^2 + \tau^2\xi^2(d\zeta)^2.$$

Thus $\quad ds = d\sigma\{1 + \tau^2\xi^2(d\zeta/d\sigma)^2\}^{\frac{1}{2}} = d\sigma\{1 + \frac{1}{2}\tau^2\xi^2(d\zeta/d\sigma)^2 - \ldots\}$

and $\qquad \dfrac{ds - d\sigma}{d\sigma} = \frac{1}{2}\tau^2\xi^2\left(\dfrac{d\zeta}{d\sigma}\right)^2 - \ldots.$

Since $(d\zeta/d\sigma)^2$ is not greater than unity, we see that, if $\tau\xi$ be a small quantity of the first order, $(ds - d\sigma)/d\sigma$ is a small quantity of the second order, for it is proportional to $\tau^2\xi^2$. Thus, when $1/\tau$ is treated as infinite in comparison with the width of the strip, we may consider that the distance between any two neighbouring points remains unchanged when the plane strip is twisted into a helicoid.

44. Stresses in a twisted blade. Let $ACA'C'$ (Fig. 17) represent the central plane of the blade before it is twisted, the faces of the blade being at a distance b on either side of this plane. When the blade is twisted, the line AOA' will, by symmetry, remain straight, and, since the twisting is uniform, every straight line initially in the plane AOC and parallel to OA will also remain straight. Thus the central plane of the blade will become a helicoid.

We have seen in § 43 that, when a strip of a mathematical plane is infinitesimally twisted into a helicoid, the distance between any two neighbouring points remains unchanged. Hence all the filaments in the *central plane* of the untwisted blade remain unchanged in length when the blade is twisted.

Let $HKLM$ (Fig. 18) be a small portion of the blade such that either face was a square before the blade was twisted, and suppose that MH and HK were initially inclined at $\pm\frac{1}{4}\pi$ to OA, as is indicated by the small square in Fig. 17.

Let I be the particle at the centre of $HKLM$ before it was twisted, so that I was midway between the faces, and let *straight lines* IU, IV be drawn through I parallel to the sides of the initial squares. Then the filaments which initially coincided with IU and IV are bent by the torsion of the blade into arcs of circles of radius $1/\tau$, the centres of curvature lying on opposite sides of the blade.

By the method employed for obtaining equation (1) of § 29, we can show that, when the blade is twisted, the filaments initially parallel to IU and at a distance h cm. from the plane IUV, measured in the *positive* direction of y, receive an elongation

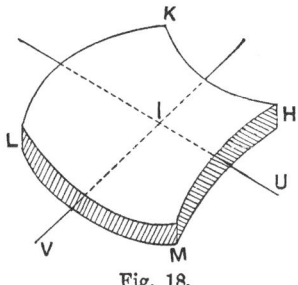

Fig. 18.

τh, while those parallel to IV receive an elongation $-\tau h$ cm. per cm. Hence, if we take an infinitesimal cube of edge q cm. with its centre at a distance h from the plane IUV and with two edges parallel to IU and IV, these edges will become $q(1+\tau h)$ and $q(1-\tau h)$ respectively.

Since the faces of the blade are free from stress, there will be no pressure on those faces of the cube which are parallel to the plane IUV. If the stresses on the other faces be a tension of R dynes per square cm. and a pressure S, as is indicated in Fig. 19, we see, by §§ 17, 18, Chapter I, that the elongations are connected with the stresses by the equations

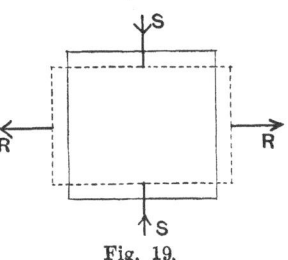

Fig. 19.

$$\tau h = \frac{R}{E} + \frac{\sigma S}{E}, \quad \tau h = \frac{\sigma R}{E} + \frac{S}{E}.$$

Hence
$$R = S = \frac{\tau h E}{1 + \sigma}.$$

But, by equation (11) of § 19, Chapter I,
$$E/(1 + \sigma) = 2n,$$

and thus
$$R = S = 2n\tau h \text{ dyne cm}^{-2}. \quad\dots\dots\dots\dots\dots(35)$$

Since $R = S$, there will be no change in those edges of the cube which are perpendicular to the plane IUV.

We shall now determine the stresses which must be applied to the edges of the blade to maintain the equilibrium of the

elements in the immediate neighbourhood of the edges.   Let
PQT (Fig. 20) be a triangular lamina of thickness
dh at a distance $+h$ from the central plane of the
blade and let the side QT lie along the edge BB'
(Fig. 17) of the blade.   Let $PQ = PT = r$ and let
QPT be a right angle.   Then PQ is acted on by a
force $R.rdh$ at right angles to PQ and PT is acted
on by a force $S.rdh$ at right angles to PT, as
shown in Fig. 20.   These forces have a resultant
$(R+S)rdh/\sqrt{2}$ parallel to QT; by (35) this is equal
to $2\sqrt{2}.n\tau hrdh$.   Since the faces of the lamina are free from
stress, a force $2\sqrt{2}.n\tau hrdh$ must be applied to the vertical edge
of the lamina in the direction QT to maintain equilibrium.   This
force is distributed over an area $r\sqrt{2}.dh$ and hence, if F be the
required tangential stress,

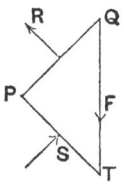

Fig. 20.

$$F = 2n\tau h \text{ dyne cm.}^{-2}.....................(36)$$

Corresponding results hold good for the remaining edges of
the blade.

45. Determination of the torsional couple.   Let Fig. 21
represent the blade seen in perspec-
tive, the thickness ($2b$) being greatly
magnified.   Let KL and MN be two
straight lines drawn on the edge
BB' perpendicular to the plane of
the blade, the distance between the
lines being dz.   Then the couple
exerted on the rectangle $KLMN$
by the tangential stress is

$$dz \int_{-b}^{+b} Fh\,dh.$$

But, by (36),

$$dz \int_{-b}^{+b} Fh\,dh = dz \int_{-b}^{+b} 2n\tau h^2 dh$$
$$= \tfrac{4}{3} n\tau b^3 dz.$$

A couple equal to this could be
produced by two horizontal forces,

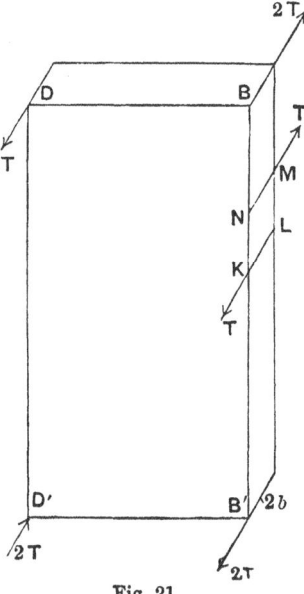

Fig. 21.

each equal to T, acting along LK and NM in opposite directions as shown in Fig. 21, if T be given by

$$T = \tfrac{4}{3}\, n\tau b^3 \text{ dynes.} \quad \ldots\ldots\ldots\ldots\ldots\ldots(37)$$

By Saint-Venant's principle (§ 24), this couple would produce the same effects as the couple arising from the vertical tangential stress F, except, of course, in the immediate neighbourhood of the edge. Since the blade is of infinitesimal thickness, the region where the effects of the two couples are appreciably different is also infinitesimal.

If we take the next element (of length dz) above MN and apply the same process to it, we shall have another pair of forces each equal to T. The lower force of this pair will act along MN in the opposite direction to the upper force of the pair corresponding to $KLMN$, and will therefore neutralise it. Proceeding in this way, we see that the forces acting on the edge BB' are equivalent to one force T applied at B in the direction away from the reader and a second force T applied at B' in the opposite direction, as is indicated in Fig. 21.

Similarly, we may replace the forces on the edge BD by one force T applied at B in the same direction as that arising from the edge BB' and another force T applied at D in the opposite direction. The total force at B is therefore $2T$.

Thus we see that the forces distributed over the four edges of the blade may be replaced by two forces each equal to $2T$ applied at B and D' away from the reader, together with two other forces of equal magnitude applied at B' and D in the opposite direction, as shewn in Fig. 21.

If the couple formed by the forces applied at B and D be G dyne-cm., we have $G = 2T \cdot BD = 2T \cdot 2a$, and thus by (37),

$$G = \tfrac{16}{3}\, n\tau a b^3.$$

If one end of the blade be twisted through ϕ radians relative to the other and if the length of the blade be l cm., we have, by (28),

$$\tau = \phi/l,$$

and thus we find

$$\frac{Gl}{n\phi} = \tfrac{16}{3}\, ab^3. \quad \ldots\ldots\ldots\ldots\ldots\ldots(38)$$

This result agrees with that given by equation (27) in § 40, when b/a is so small that $3\cdot361b/a$ is negligible compared with $16/3$

In experimental work it will not generally be possible to apply forces at B, B', D and D' in the manner shown in Fig. 21. But a second application of Saint-Venant's principle leads to the conclusion that, provided they be equivalent to a couple G, the manner in which the forces are distributed along BD or along $B'D'$ is of no consequence except at points near the ends of the blade. Any uncertainty due to this cause is small when the length of the blade is great compared with its width.

A model of a helicoid may be made of a strip of paper about one centimetre in width, which is kept tight while it is twisted. If the lines in the diagrams be drawn upon the strip the reader will be aided in following the discussion above.

CHAPTER III.

EXPERIMENTAL WORK IN ELASTICITY.

46. Introduction. In this chapter descriptions are given of a number of experimental methods of studying the elastic properties of solid bodies. Most of the experiments are directed towards obtaining values of Young's modulus or of the rigidity, or the ratio of one of these moduli to the other, and here the strains are assumed to be so small that Hooke's law is obeyed accurately. In other experiments, the deviations from Hooke's law and their effects are studied.

Though, in nearly every case, the apparatus is so simple that it may be constructed by any person who is moderately skilled in the use of tools, yet the experimental methods, when carried out with care, are capable of yielding definite results. The word *definite* is used here to imply that, when a determination of an elastic quantity has been made, a repetition of the experiment upon the *same* specimen and under the *same* conditions will lead to a result which does not differ by more than one or two per cent. from that of the first determination. The impossibility, in most cases, of securing truly homogeneous and isotropic material makes it useless to expect that the value of the elastic quantity deduced from the experiment will be anything more than a rough sort of average value*. In some cases the observations are taken on scales divided to millimetres, and the necessity of keeping within Hooke's law often limits the measured displacement to one or two centimetres. It is clear that, in these cases, very careful readings are required if the result is to be accurate to within two or three per cent.

---

* See the last paragraph of § 39.

When the result depends upon the fourth power of the radius of a wire, particular attention should be given to measuring the diameter of the wire with a screw-gauge, since an error of one per cent. in the radius involves an error of four per cent. in the result*.

The work will gain considerably in interest if the student is able to test the same specimen by different methods. Thus, for a given piece of wire, Young's modulus may be found as in EXPERIMENT 2 or 3, the rigidity may be found as in EXPERIMENT 4 or 5, while Young's modulus may be found for a portion of the wire as in EXPERIMENT 7 or even as in EXPERIMENT 6 or 10, if the distance between the knife edges be small and the loads light. Similarly, the same rod may be used for EXPERIMENTS 6 and 10.

EXPERIMENT 1.　**Experimental Investigation of Hooke's Law for Copper.**

47. Introduction. The mathematical theory of elasticity is based upon the assumption that, for a given stress the strain is independent of the time and that, for small strains, stress and strain are proportional so that, in Hooke's words, *Ut tensio sic vis.* We shall therefore begin the experimental part of the subject by describing a sensitive method of investigating the relation between stress and strain in a wire subject to small elongations. Young's modulus, if it exist, i.e. if Hooke's law hold, can be found with sufficient accuracy by the apparatus described in § 52. The object of the present experiments is not so much to obtain a very accurate value for Young's modulus as to gain a working knowledge of the natural habits of the material under test.

The simplest method of magnifying the effects to be observed consists in using a wire of considerable length, hung from a beam or other support, the extension being produced by hanging weights to the lower end of the wire; but this method is liable to two serious errors. These arise from the yielding of the support and from the change of length of the wire due to rise of temperature.

---

* Many screw-gauges have the defect that the pitch of the screw, i.e. the distance it advances for one revolution, is not clearly marked on the instrument. In such cases the student should ascertain the pitch from the teacher, or, in a practical examination, from the examiner.

The latter cause may introduce a comparatively large error, for a rise of 2° C. will produce an increase of length of $\frac{1}{10}$ mm. in a copper wire 3 metres long.

The first error is eliminated if, instead of finding the displacement of the lower end of the wire relative to a fixed mark, we observe the displacement of the end of the wire relative to the lower end of a second wire of the same material, hanging from the same support, stretched by a constant weight and serving as a standard for comparison.

The use of a comparison wire practically eliminates the second error also, for it is found that the coefficient of linear expansion of a wire is but little affected by variations in the load carried by the wire. Thus, Dr J. T. Bottomley (*Phil. Mag.* 1889, Vol. 28, p. 94) made experiments on a pair of copper wires 3.8×10^{-4} cm.^{2} in section, one being stretched by a load of 375 grammes and the other by a load of 75 grammes. Dr Bottomley measured directly, by means of a sensitive mirror method, the excess of the extension of the first over that of the second wire, when they were heated simultaneously, and found the coefficient of *relative* expansion to be 3.14×10^{-7} degree^{-1}. Since the difference of tension was

$$300 \times 981/(3.8 \times 10^{-4}) \text{ or } 7.74 \times 10^8 \text{ dyne cm.}^{-2},$$

we find that, if the coefficient of linear expansion of a copper wire be α degree^{-1} and if the tension of the wire be T dyne cm.$^{-2}$, then

$$\frac{d\alpha}{dT} = \frac{3.14 \times 10^{-7}}{7.74 \times 10^8} = 4.05 \times 10^{-16} \text{ degree}^{-1} \text{ dyne}^{-1} \text{ cm.}^2.$$

The value of α for copper is about 1.72×10^{-5} degree^{-1}, so that the extra load of 300 grammes increased the coefficient of expansion by one part in 55.

If two copper wires, each one square mm. in section, carry loads differing by one kilogramme, the difference of tension will be about 10^8 dynes cm.$^{-2}$, and hence if the wires be 3 metres long, a rise of 1° C. will cause the wire with the greater load to extend by $4.05 \times 10^{-8} \times 300$ or by 1.22×10^{-5} cm. more than the other wire. This difference of extension is too small to be measurable with the apparatus described in § 48.

The value of $d\alpha/dT$ may also be deduced from the temperature coefficient of Young's modulus (E) for a wire with a constant tension.

If l_0 be the length of a wire at $0°$ C. and under zero tension, and if l be its length when the temperature is $\theta°$ and the tension is T, l/l_0 is a function of T and θ, and thus we have

$$\frac{d}{dT}\left(\frac{1}{l_0}\frac{dl}{d\theta}\right) = \frac{d}{d\theta}\left(\frac{1}{l_0}\frac{dl}{dT}\right)$$

or, by Chapter I, § 17, $\dfrac{d\alpha}{dT} = \dfrac{d}{d\theta}\left(\dfrac{1}{E}\right) = -\dfrac{1}{E^2}\dfrac{dE}{d\theta}$.

Mr G. A. Shakespear has found (*Phil. Mag.* 1889, Vol. 47, p. 551) the values of $E^{-1}dE/d\theta$ given in the first column of the table. Assuming that E had the values given in the second column, we obtain, by the last equation, the values of $d\alpha/dT$ given in the third column.

Metal	$\dfrac{1}{E}\cdot\dfrac{dE}{d\theta}$	E	$\dfrac{d\alpha}{dT}$
Copper ...	$-4\cdot1 \times 10^{-4}$	$1\cdot2 \times 10^{12}$	$3\cdot4 \times 10^{-16}$
Hard Brass ...	$-3\cdot4 \times 10^{-4}$	$1\cdot0 \times 10^{12}$	$3\cdot4 \times 10^{-16}$
Soft Iron ...	$-1\cdot8 \times 10^{-4}$	$2\cdot0 \times 10^{12}$	$0\cdot9 \times 10^{-16}$
"Silver" Steel	$-3\cdot7 \times 10^{-4}$	$2\cdot0 \times 10^{12}$	$1\cdot8 \times 10^{-16}$

On account of the difficulty of the experiments, the agreement between the values of $d\alpha/dT$ for copper obtained by the two methods is perhaps as close as could be expected.

48. Apparatus. By the instrument shown in Fig. 22 very small extensions of the wire under test can be measured relatively to the comparison wire*.

The two wires A, A', have their upper ends secured to a stout piece of metal bolted to a beam. From the lower ends hang two brass frames CD, $C'D'$, supporting the two ends of a sensitive level L. One end of the level is pivoted to the frame CD by the pivots H; the other end of the level rests upon the end of a vertical screw S working in a nut attached to the frame $C'D'$. The two links K, K' prevent the frames from twisting relatively to

---

* G. F. C. Searle, *Proc. Cambridge Phil. Soc.* Vol. x, p. 318 (1900).

each other about vertical axes, but freely allow vertical relative motion. When these links are horizontal, the two wires are parallel to each other. From the lower ends of the frames CD, $C'D'$, hang a mass M and a pan P represented diagrammatically in the figure. The weights of M and P are sufficient to ensure

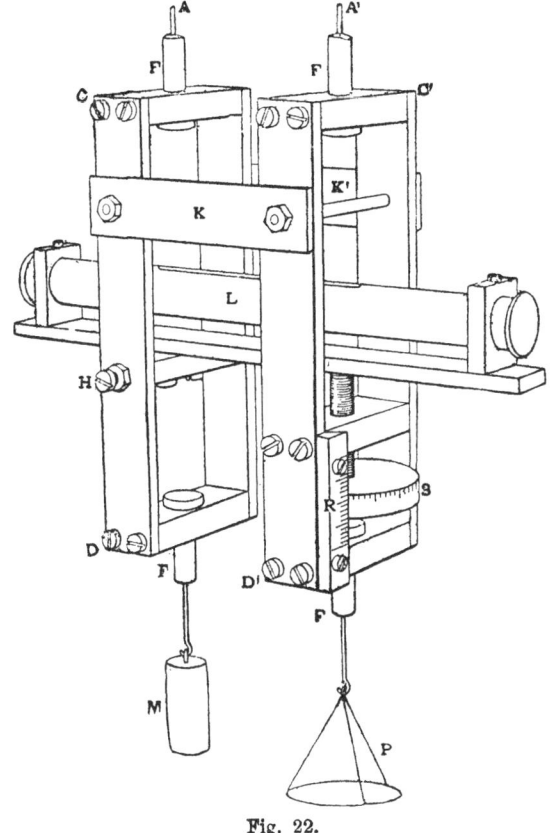

Fig. 22.

that the wires are straight. The connexions between the wires and the frames are made by the swivels F into which the ends of the wires are soldered. The swivels enable the observer to set the wires free from torsion and thus to ensure that the two wires hang in a vertical plane. Two other swivels connect M and P to the frames.

The head of the screw is divided, and a scale R engraved on the side of the frame serves to determine the number of complete revolutions made by the screw. In the instrument in use at the Cavendish Laboratory the pitch of the screw is $\frac{1}{2}$ mm., while the head is divided into 100 parts. Each division on the head thus corresponds to $\frac{1}{200}$ mm. A pitch of 1 mm. would be more convenient*.

The instrument is used in the following manner. Suppose that the screw has been adjusted so that one end of the bubble of the level is at its fiducial mark†. A fine wire passing round the level and held tight by an indiarubber band may be used as a fiducial mark. If the wire be arranged to be in the plane of two of the vertical sides of one of the frames, errors of parallax can be avoided by taking observations with the eye in this plane. If a mass be placed in the pan P, the wire A' is stretched and the bubble moves towards H. The bubble is then brought back to its fiducial mark by turning the screw so as to raise the end of the level resting upon it. The distance through which the screw is moved is clearly equal to the increase of length of the wire A' and is determined at once by the difference of the readings of the screw in the two positions. In the Cavendish Laboratory instrument, the level is sensitive enough to enable the screw to be adjusted to $\frac{1}{5}$ of a division on its head, i.e. to $\frac{1}{1000}$ mm.

To steady the instrument, it is convenient to allow the two wires to press lightly against a rod fixed horizontally at a small distance above the frames CD, $C'D'$.

A brief discussion of the kinematics of the instrument may be added. In order to secure that there shall be only one possible displacement of one frame relative to the other, five out of the six degrees of relative freedom must be destroyed. Since only relative motion is in question, we may imagine one frame, say CD, to be fixed. The other frame $C'D'$ is kept vertical by the tensions of the

* The chief dimensions of the apparatus are as follows:—$CD = 11$ cm. Length of links 5 cm. Diameter of screw head $= 4$ cm.

† If the glass tube of a level be not well secured in the metal tube which protects it, an attendant, in cleaning the apparatus, may cause a rotation of the glass tube about its axis. If this cause the ends of the tube to be higher than the centre, it will be impossible to adjust the supports of the level so as to bring the bubble to the centre; the bubble will always go to one end or the other.

wires above and below it. If the links were absent, it would be free to move horizontally East and West, or North and South, or vertically (when the wire is stretched) and to turn about a vertical axis. It would thus possess four degrees of freedom. The two links destroy three degrees of freedom by preventing the frame from (1) rotating about its own wire A', (2) moving horizontally at right angles to the links, (3) moving towards or away from the frame CD. The frame $C'D'$ has thus but one degree of freedom remaining, viz. that which enables it to follow the stretching of the suspending wire.

49. Experiments on loading and unloading a copper wire. One of the most interesting uses of the instrument is to find the changes of length of a copper wire, which occur when the load in the pan is increased step by step from zero to any value W and is then diminished to zero again. When the load is changed, the wire only gradually assumes the length corresponding to the new load, and thus the readings will gain in regularity if the changes of load be made at approximately equal intervals—say of two minutes. The observations may be made in the following manner. Starting with the pan empty, a reading of the micrometer is taken and is *recorded*. A mass is then placed in the pan and after two minutes (or whatever interval is chosen) the reading of the micrometer is again recorded, and the process is continued with equal steps in the load till the maximum load W is reached; the load is then reduced step by step to zero. The masses should be put in and taken out of the pan as gently as possible.

When the *initial* micrometer reading for zero load is subtracted from the other readings the differences are the extensions of the wire. The results are plotted on squared paper, the abscissae representing loads and the ordinates extensions.

When the wire has been unloaded for a comparatively long time before the initial reading for zero load is taken, a curve similar to that in Fig. 23 will be obtained. In this case the wire has been loaded during the whole cycle of observations and, in consequence, at the end of the cycle the wire is longer by a few thousandths of a millimetre than it was at the beginning.

If, on the other hand, a load W_0, at least as great as W, be placed in the pan for a comparatively long time and if the load be removed only a short time before the initial reading for zero load is taken, the curve representing the results of loading and unloading will be similar to that in Fig. 24, the wire being slightly *shorter* at the end than at the beginning of the cycle. The explanation lies in the fact that during the later stages of the second half of the cycle the load is less than W_0, and thus the wire has had opportunity and time to contract a little.

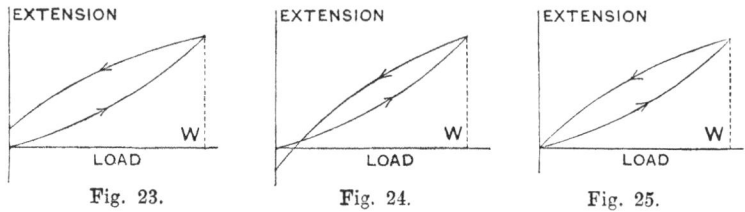

Fig. 23.   Fig. 24.   Fig. 25.

If, by repeatedly loading and unloading it, the wire could be brought to a thoroughly cyclic state, the curve would be as in Fig. 25, the final and initial readings being identical. But the establishment of the cyclic state would occupy much time, since each cycle of loading and unloading would have to be made *at the same rate* as the cycle during which the readings are taken. If the preliminary cycles are made comparatively rapidly the curve will be as in Fig. 24.

50. Graphical representation of deviations from Hooke's law. Hooke's law is so nearly true for stresses, which are small compared with the breaking stress, that it is impossible to exhibit in a satisfactory manner both the whole extension and also any deviation from Hooke's law on the same diagram. We may, however, adopt a device which is useful whenever small deviations disturb the strict proportionality between cause and effect. If the maximum load W produce an extension Z, and if z be the extension due to any smaller load w, we subtract wZ/W from z and denote $z - wZ/W$ by d. We then plot the difference d against the load w. If Hooke's law were exactly fulfilled, the difference would vanish for every value of w, and thus these differences show the departures from Hooke's law. This method has been adopted in the diagram given in § 51.

51. Practical example. The observations may be entered as in the following record of an experiment by Mr Field upon a copper wire 285·7 cm. in length and about 0·0119 cm.^{2} in cross section. To save space, only the extensions are entered below; but the student should *record* the reading of

$W=6$ kilo. $Z=0·1159$ cm. $Z/W=0·01932$			$W=4$ kilo. $Z=0·0763$ cm. $Z/W=0·01908$			$W=2$ kilo. $Z=0·0371$ cm. $Z/W=0·01855$		
w kilo.	z cm.	$1000\,d$ cm.	w kilo.	z cm.	$1000\,d$ cm.	w kilo.	z cm.	$1000\,d$ cm.
0	·0000	0·0	0	·0000	0·0	0	·0000	0·0
1	·0184	− 0·9	1	·0186	− 0·5	1	·0184	− 0·2
2	·0375	− 1·1	2	·0376	− 0·6	2	·0371	0·0
3	·0567	− 1·3	3	·0568	− 0·4	1	·0186	0·0
4	·0763	− 1·0	4	·0763	0·0	0	− ·0001	− 0·1
5	·0964	− 0·2	3	·0581	+ 0·9			
6	·1159	0·0	2	·0389	+ 0·8			
5	·0978	+ 1·2	1	·0192	+ 0·2			
4	·0792	+ 2·0	0	− ·0003	− 0·3			
3	·0598	+ 1·8						
2	·0404	+ 1·8						
1	·0207	+ 1·4						
0	− ·0032	− 3·2						

the micrometer for every load and then deduce the extensions from those readings. In the tables the load w and the extension z are given for cycles of loading and unloading. The maximum load is W, and the maximum extension is Z. The quantity $d = z - wZ/W$ shows the departure from Hooke's law.

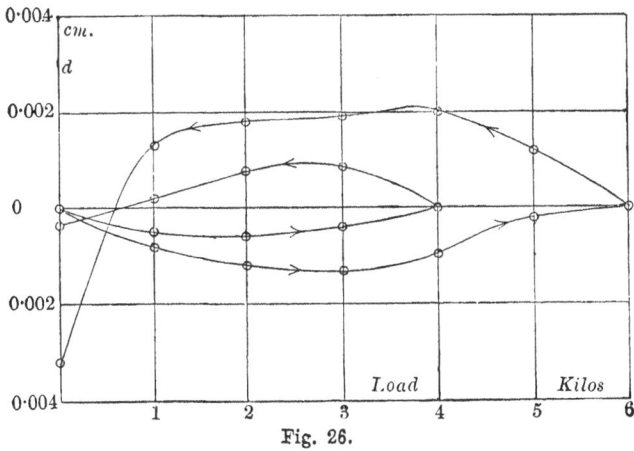

Fig. 26.

The results for $W=6$ and for $W=4$ kilogrammes are shown in Fig. 26. An attempt was made to reduce the wire to a cyclic state, but the curves show that the attempt failed, the curves resembling that of Fig. 24. Very careful work is necessary to obtain curves as regular as those obtained by Mr Field.

EXPERIMENT 2. **Determination of Young's modulus by stretching a vertical wire.**

52. Apparatus. The simplest method of determining Young's modulus depends upon observations of the increase of length of a long vertical wire when the load carried by it is increased. Since Hooke's law begins to fail when the elongation is much more than $\frac{1}{1000}$ cm. per cm., the wire should be of considerable length, so that, without going beyond the elastic limit, the increase of length may be large enough for satisfactory observation. When the changes of length are observed by means of a millimetre scale fitted with a vernier reading to $\frac{1}{10}$ mm., errors of one per cent. will probably occur in the measurements unless the changes exceed one cm. Hence, if the elongation is not to exceed $\frac{1}{1000}$ cm. per cm., the wire should be at least 10 metres long. When a more sensitive appliance, such as that described in § 48, is available for measuring the change of length, satisfactory results can be obtained with comparatively short wires.

Since the elongation is small, it is necessary to take special precautions against two sources of error. These arise from the yielding of the support and the change of length of the wire due to a change of temperature during the experiment. Both errors are practically eliminated if, instead of finding the displacement of the lower end of the wire relative to a *fixed* mark, we observe its displacement relative to the lower end of a second wire of the same material, hanging from the same support and carrying a constant load. This measurement is easily made if a scale be attached to the end of one wire and a vernier to the end of the other. Any yielding of the support affects both wires equally, and any change of temperature causes very nearly the same expansion in both wires in spite of the difference between the loads, since, as is shown in § 47, the coefficient of expansion depends only very slightly upon the load carried by the wire.

The two wires are secured to a block of metal attached to

a beam or other firm support in the manner shown in Fig. 27. A millimetre scale is clamped to the comparison wire, which carries the constant load, and a vernier reading to $\frac{1}{10}$ or $\frac{1}{20}$ mm. is clamped to the wire which is to be stretched. The vernier is kept in the proper position against the scale by a **V**-slide or by simple guides attached to the vernier. A scale pan, not shown in the figure, is hung from the wire below the vernier and a constant load is hung from the comparison wire below the scale. The scale pan and the constant load must be heavy enough to ensure that the wires are straight.

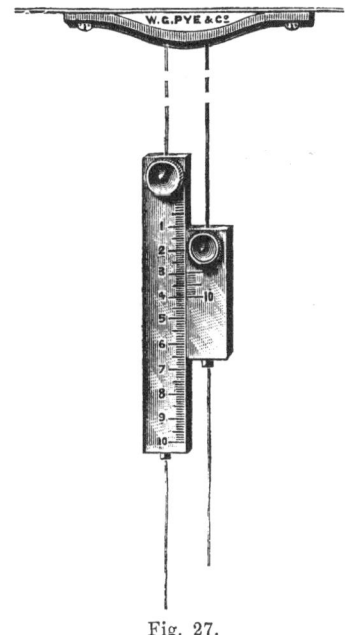

Fig. 27.

53. Determination of Young's modulus. In taking the observations the vernier is first read with no additional mass in the scale pan. The load is then increased by steps of 1 kilogramme up to, say, 6 or 8 kilogrammes and the vernier is read at each stage. The load is then diminished step by step and the vernier is again read at each stage. The masses must be put into the pan carefully, so as to avoid the great increase of stress which occurs when a mass is allowed to *drop* into the pan*. If the reading of the vernier at the end of this cycle of operations differ appreciably from the reading at the beginning, the wire has been permanently stretched, and the observations cannot be used for finding Young's modulus. In this case a new set of observations must be made with a smaller maximum load, not great enough to give any appreciable permanent set†.

Trouble due to overloading the wire will be practically avoided if the maximum load be not allowed to exceed half the breaking load. If the cross-section of the wire be A cm.², and M_{max} grammes be the maximum load, then M_{max} may be calculated by the formula

$$M_{max} = \tfrac{1}{2}A \cdot B/981,$$

where B dyne cm.⁻² is the breaking stress or the tenacity. Rough values of B are given in the table.

Material	B
Brass	3×10^9 dyne cm.⁻²
Copper	4×10^9
Iron	6×10^9
Steel	10×10^9

When a satisfactory set of observations has been obtained, the mean of the two readings of the vernier for each load is subtracted from the mean reading for the empty pan, the difference in each case being the mean extension due to the corresponding load.

---

* Let the length of the wire change from L_0 to $L_0 + l$ when the load is *gradually* increased from M_0 to $M_0 + M$. The effect of *suddenly* applying M may be illustrated by supposing that M is suspended by a string so as to just touch the pan and that the string is cut. The mass $M_0 + M$ will then oscillate about a mean position in which the length of the wire is $L_0 + l$. At the highest point of the oscillation the length is L_0 and, therefore, at the lowest point the length is $L_0 + 2l$. Hence the maximum increase of tension is twice that due to a gradual increase of load from M_0 to $M_0 + M$. If M had been allowed to *drop* into the pan the effect would have been greater.

† If the wire has been *freshly* set up, the first addition of a considerable load may permanently change the reading of the vernier for the pan alone by straightening out kinks in the wire.

The readings may be recorded as in § 54, and should be expressed in *centimetres*.

The mass of the pan does not cause any difficulty as long as Hooke's law holds. For suppose that the mass of the pan is M_0 grammes and that it produces an extension l_0 cm., and that an additional load M grammes increases the extension by l cm. Then we have, by Hooke's law,

$$\frac{M_0 + M}{l_0 + l} = \frac{M_0}{l_0} = C, \text{ say.}$$

Hence
$$\frac{M}{l} = C.$$

Thus the ratio of the added load to the increase of extension due to that load is the same as the ratio of the whole load to the whole extension, and therefore, in finding Young's modulus, we may neglect entirely the mass of the pan and the extension due to it.

If the values of l/M prove to be nearly constant for different loads, the mean value of l/M may be used in finding Young's modulus. When the irregularities are serious, the results should be shown graphically on squared paper, the ordinate representing the mean extension due to each added load, while the abscissa represents the added load. Since Hooke's law is assumed to hold, a straight line should be drawn, by the aid of a stretched thread, so as to lie as evenly as possible among the points plotted on the diagram. When the best position of the thread has been found, it is recorded by two marks made on the paper, one near each end of the thread. These marks are then joined by a line drawn by the aid of a *straight* ruler; many wooden scales are far from straight. The difference between the ordinates of two points *on this line* corresponding to $M = 0$ and to some definite mass M (say 5 kilogrammes) is taken as the value of l for that mass. The corresponding value of M/l is used in calculating Young's modulus.

The length of the wire from the point of support to the clamp, which fixes the vernier to the wire, may be determined by a tape measure or by the aid of a long rod which is afterwards measured by one or more metre scales. This length should be expressed in centimetres. During these measurements the wire should be kept straight by the weight of the pan. Strictly speaking, the initial length should be measured when the wire carries no load, but, since the increase in the length of the wire due to the pan alone will perhaps not exceed one part in 10,000, the length of the wire when carrying the pan alone may be taken as equal to L centimetres, the initial length of the wire.

It should be noted that L and l are obtained by two distinct sets of measurements. In finding the length by the tape measure we are not concerned with the extension, and in finding the extension by the vernier we are not concerned with the length of the wire. It would be difficult to measure a length of, say, 5 metres to within 2 mm. by a tape measure, but, in view of other uncertainties, an error of 2 mm. in the length of the wire may fairly be neglected. Yet an error of 2 mm. in the determination of the extension would render the results worthless.

To complete the measurements, the cross section of the wire must be obtained. If the wire be permanently fixed to the support, the diameter is found by a screw-gauge. Readings are taken at 4 or 5 points on the wire between the support and the vernier, and two diameters at right angles are measured at each point, care being taken not to compress the wire in taking the readings. The zero reading of the screw-gauge is observed, and the corresponding correction is applied to the readings, which should be expressed in *centimetres*. The mean radius*, a, is found by halving the mean diameter and then the cross section A is calculated in square cm. from the expression $A = \pi a^2$†.

If the wire can be removed from the support, the volume of the part between the support and the vernier can be found by the hydrostatic balance. If this be V c.c., then $A = V/L$ square cm.

When the load is M grammes, the longitudinal stress T is Mg/A dyne cm.$^{-2}$. If this load correspond to an increase of length of l cm. in a total length of L cm. the elongation e, i.e. the

---

* The quantity we are really concerned with is not the mean radius but the square root of the reciprocal of the mean value of (radius)$^{-2}$. The appropriate correction is calculated in Note VI, § 1.

† The neglect of the distinction between the radius and the diameter of a wire is a frequent cause of disaster in students' work.

increase of length per unit length, is l/L cm. per cm.   Hence, by
Chapter I, § 17, Young's modulus is given by

$$E = \frac{\text{stress}}{\text{elongation}} = \frac{T}{e} = \frac{Mg/A}{l/L} = \frac{MgL}{lA} \text{ dynes per square cm.}$$

54. Practical example. The observations may be entered as in the
following record of an experiment made on a brass wire.

Length of wire from support to clamp of vernier $= L = 745$ cm.

Readings of screw-gauge on wire

$$\left.\begin{array}{llll} 0\cdot0944 & 0\cdot0943 & 0\cdot0947 & 0\cdot0945 \\ 0\cdot0943 & 0\cdot0943 & 0\cdot0945 & 0\cdot0945 \end{array}\right\} \text{ mean reading } 0\cdot0944 \text{ cm.}$$

Correction for zero $0\cdot0002$ cm., to be added.

Hence diameter $= 2a = 0\cdot0946$ cm.

Cross section $= A = \pi a^2 = \pi \times (0\cdot0473)^2 = 0\cdot00703$ square cm.

Readings of vernier for increasing and diminishing loads :—

Load grammes	Loading cm.	Unloading cm.	Mean cm.	Extension cm.	$\dfrac{\text{Extension}}{\text{Load}}$ cm. grm.$^{-1}$
0	1·510	1·515	1·513	0	—
1000	1·625	1·625	1·625	·112	$1\cdot12 \times 10^{-4}$
2000	1·730	1·735	1·733	·220	1·10
3000	1·835	1·840	1·838	·325	1·08
4000	1·945	1·945	1·945	·432	1·08
5000	2·045	2·050	2·048	·535	1·07
6000	2·150	2·155	2·153	·640	1·07
7000	2·260	2·260	2·260	·747	1·07
8000	2·355	2·360	2·358	·845	1·06
9000	2·465	—	2·465	·952	1·06

Mean $1\cdot079 \times 10^{-4}$

When the extension was plotted against the load, it was found that the
points lay very nearly on the straight line cutting the line $M=0$ at $0\cdot006$ cm.
and the line $M=5000$ at $0\cdot533$ cm.   Hence, for $M=5000$ grm., $l=0\cdot527$ cm.
The corresponding value of l/M, viz. $0\cdot527/5000$ or $1\cdot054 \times 10^{-4}$ differs by
2·5 per cent. from the mean value of l/M derived from the table.   The zero
reading $1\cdot513$ cm. in the table is clearly abnormal; possibly the weight of
the empty pan is insufficient to ensure that the wire is straight.   If we treat
the reading $1\cdot733$ cm., which was found for a load of 2000 grammes, as the
zero reading, we find the following values of l/M :—

1·050,   1·060,   1·050,   1·050,   1·054,   1·042, $1\cdot046 \times 10^{-4}$.

The mean, $1\cdot050 \times 10^{-4}$, now agrees closely with the value derived from the diagram. Using the value $1\cdot054 \times 10^{-4}$ for l/M, we have, by § 53,

$$\text{Young's modulus} = E = \frac{MgL}{lA} = \frac{981 \times 745}{1\cdot054 \times 10^{-4} \times 0\cdot00703}$$
$$= 9\cdot86 \times 10^{11} \text{ dynes per square cm.}$$

EXPERIMENT 3. **Determination of Young's modulus by stretching a horizontal wire.**

55. Apparatus. One end of a wire, 1 to 2 metres in length, is soldered or otherwise secured to a block of metal B (Fig. 28)

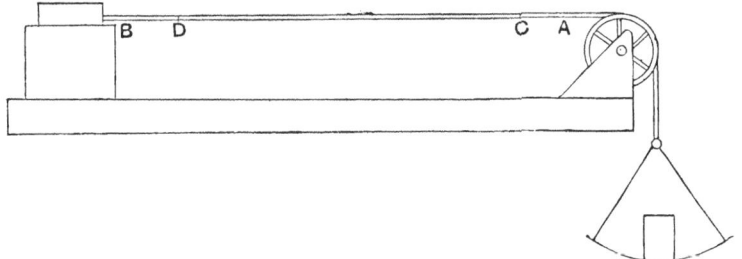

Fig. 28.

which is firmly clamped to a table. The wire passes over a pulley A and is stretched by weights placed in a pan. This pan must be made heavy enough, by the use of permanent weights, to prevent the wire from sagging appreciably when no additional weights are in the pan (see § 57). Two scratches, C and D, are made on the wire near the pulley A and the block B, and the movements of these scratches are observed by means of two travelling microscopes, which are focussed on the wire.

If travelling microscopes are not available, microscopes with micrometer scales in their eyepieces must be used. The value of one division of the micrometer scale of each microscope is deduced from the number of micrometer divisions covered by the image of one division of a millimetre scale. The dividing lines of this millimetre scale must be *fine*; if they are coarse, it will be impossible to obtain an accurate value for the micrometer divisions.

It is, however, not necessary to use two microscopes if it is found, on examination by a microscope, that the block B does not move appreciably when the pan is loaded. In this case we may regard D as coinciding with the end of the block B.

56. Determination of Young's modulus. The load in the pan is increased by equal steps of 500 or 1000 grammes from zero up to some maximum value, and is then diminished by equal steps, and at each stage the readings of the scratches C and D are taken by the microscopes. If a load of M grammes placed in the pan cause C and D to move through x cm. and y cm. from their positions for zero load in the pan, the increase in the length of CD due to M is $x - y$ cm.

If the value of $(x - y)/M$ prove to be nearly constant for different loads, the mean value may be used in finding Young's modulus. If there are serious irregularities, the results should be shown on squared paper, the abscissa representing M, while the ordinate represents the *mean* of the corresponding values of $x - y$ for increasing and diminishing loads. A straight line is drawn by the aid of a stretched thread, as in § 53, so as to pass as evenly as possible among the points on the diagram, and from this straight line the values of $x - y$ for $M = 0$ and for some definite load M (say 5 kilogrammes) are read off. The value of $M/(x - y)$, corrected in this way, is used in calculating Young's modulus.

The length of CD for zero load in the pan is denoted by L cm. and is obtained by means of metre scales placed end to end, and the diameter of the wire is found with a screw-gauge, corrected for zero error, at 4 or 5 points between C and D, two perpendicular diameters being measured at each point. If the mean radius, i.e. half the mean diameter, be a cm. and if the cross-section be A square cm., then $A = \pi a^2$*.

The stress T due to a load of M grammes is Mg/A dynes per square cm., and this produces an increase of $x - y$ cm. in a length of L cm.; thus the elongation e is $(x - y)/L$ cm. per cm. Hence, by Chapter I, § 17, Young's modulus is given by

$$E = \frac{\text{stress}}{\text{elongation}} = \frac{T}{e} = \frac{Mg/A}{(x-y)/L} = \frac{MgL}{A(x-y)} \text{ dynes per square cm.}$$

The observations may be tabulated as in § 58, the readings of the microscopes and the two values of $x - y$ being recorded for each value of M.

---

* See the first Footnote on page 84.

57. Notes on the method. In this experiment any error due to a possible motion of the block B is eliminated by using a microscope to observe the motion of the scratch D. There is, however, no temperature compensation, as in EXPERIMENT 2, and therefore a thermometer should be placed near the wire to give warning of any serious changes of temperature.

By the microscopes the changes of length can easily be found to within $\frac{1}{1000}$ cm., and thus the wire may be much shorter than in EXPERIMENT 2, where a vernier, reading to $\frac{1}{100}$ or $\frac{1}{200}$ cm. is used.

It is essential in this experiment that the stretching force should be always great enough to ensure that S, the length of the wire from C to D, *measured along the wire*, should not differ appreciably from L, the distance from C to D measured along a straight line. An approximate estimate of $S - L$ is easily made. Let the tangents to the wire at C and D (Fig. 29) make angles θ_1

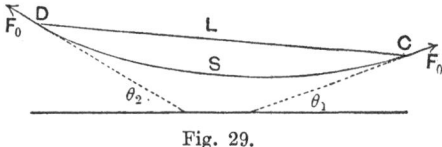

Fig. 29.

and θ_2 with the plane of the horizon, and let m be the mass of CD. If F_0 be the stretching force when the pan is empty, we may take F_0 as constant at all points of CD. Since the weight of CD is supported by the forces at C and D, we have

$$F_0 (\sin \theta_1 + \sin \theta_2) = mg.$$

If ϕ be the mean of θ_1 and θ_2, we have, since θ_1 and θ_2 are very small,

$$2F_0 \phi = mg.$$

Now, when S is small compared with the radius of curvature of the arc at its lowest point, we may treat the curve as an arc of a circle subtending an angle 2ϕ at its centre. Then, if ρ be the radius of curvature,

$$\frac{L}{S} = \frac{2\rho \sin \phi}{2\rho \phi} = \frac{\sin \phi}{\phi} = 1 - \frac{\phi^2}{6} + \cdots,$$

and thus, approximately,

$$\frac{S-L}{S} = \frac{\phi^2}{6} = \frac{m^2 g^2}{24 F_0^2}.$$

Hence, if the mass of the pan be 2 kilogrammes, so that F_0 is $2000g$ dynes, and if the mass of CD be 10 grammes,

$$(S-L)/S = 1/960,000.$$

The elongation due to the stretching of the wire in the determination of Young's modulus may be as great as $\frac{1}{1000}$ cm. per cm., and thus, since the apparent elongation due to changes of sagging when additional masses are placed in the pan does not exceed about $\frac{1}{1,000,000}$ cm. per cm., any error due to sagging may be neglected.

58. Practical example. The observations may be entered as in the following record of an experiment made by Mr T. G. Bedford upon a brass wire.

Length of wire between scratches = $L = 124\cdot4$ cm.

Readings of screw-gauge for pairs of diameters at right angles,

| ·0692 | ·0698 | ·0693 | ·0695 | ·0694 | mean reading 0·0695 cm. |
| ·0698 | ·0692 | ·0696 | ·0696 | ·0697 | |

Correction for zero error 0·0005 cm., to be added.

Diameter of wire $= 2a = 0\cdot0700$ cm.

Cross section of wire $= A = \pi a^2 = \pi (0\cdot035)^2 = 3\cdot849 \times 10^{-3}$ cm.^{2}.

By means of permanent weights the mass of the pan was made about 5·5 kilogrammes, which was sufficient to prevent any appreciable sagging of the wire.

Load M grammes	LEFT MICROSCOPE			RIGHT MICROSCOPE		
	Loading cm.	Unloading cm.	Mean cm.	Loading cm.	Unloading cm.	Mean cm.
0	·7282	·7278	·7280	·9732	·9732	·9732
1000	·7666	·7683	·7674	·9668	·9658	·9663
2000	·8054	·8091	·8072	·9594	·9582	·9588
3000	·8460	·8488	·8474	·9525	·9515	·9520
4000	·8858	·8896	·8877	·9456	·9448	·9452
5000	·9265	—	·9265	·9380	—	·9380

Two microscopes were used, each reading to $\frac{1}{200}$ mm.; they formed a "pair" and the scales were numbered in opposite directions. In the above table the readings have been reduced to centimetres, but Mr Bedford recorded the actual readings on the micrometer heads in each case. Thus the two readings, which appear in the table as ·07282 and 0·9732 cm., were recorded as $7·0 + 56·4/200$ mm. and $9·5 + 46·3/200$ mm.

From this table we obtain the following values of x, the displacement of the scratch observed by the left-hand microscope, and of y, the displacement of the other scratch.

M grammes	x cm.	y cm.	$x - y$ cm.	$(x - y)/M$ cm. grm.$^{-1}$
1000	·0394	·0069	·0325	$3·25 \times 10^{-5}$
2000	·0792	·0144	·0648	$3·24 \times 10^{-5}$
3000	·1194	·0212	·0982	$3·27 \times 10^{-5}$
4000	·1597	·0280	·1317	$3·29 \times 10^{-5}$
5000	·1985	·0352	·1633	$3·27 \times 10^{-5}$

Mean $3·264 \times 10^{-5}$

Hence, by § 56, since the mean value of $(x - y)/M$ is

$$3·264 \times 10^{-5} \text{ cm. grm.}^{-1},$$

$$\text{Young's Modulus} = E = \frac{MgL}{A\,(x - y)} = \frac{981 \times 124·4}{3·849 \times 10^{-3} \times 3·264 \times 10^{-5}}$$

$$= 9·71 \times 10^{11} \text{ dynes per square cm.}$$

EXPERIMENT 4. **Determination of rigidity. Statical method.**

59. Apparatus. In the statical method the couple is applied to the rod by means of a mass supported by a tape wound round a wheel A (Fig. 30). The wheel is fixed to a steel axle B, supported by the bearing C. The rod to be tested may be about 0·4 cm. in diameter and 45 cm. in length and should be as straight as possible. One end of the rod is attached to the axle and the other end is fixed to a block D, both the bearing C and the block D being firmly secured to a stout base board E. One end of a thin tape is attached to the wheel and the other end to a pan P, care being taken that the vertical portion of the tape is always tangential to the wheel.

The rod may be connected with the axle by a block of brass, into which both the rod and the axle are soldered, or the rod may be gripped by a self-centering three-jawed chuck attached to the axle, the latter plan being convenient when more than one rod is to be tested.

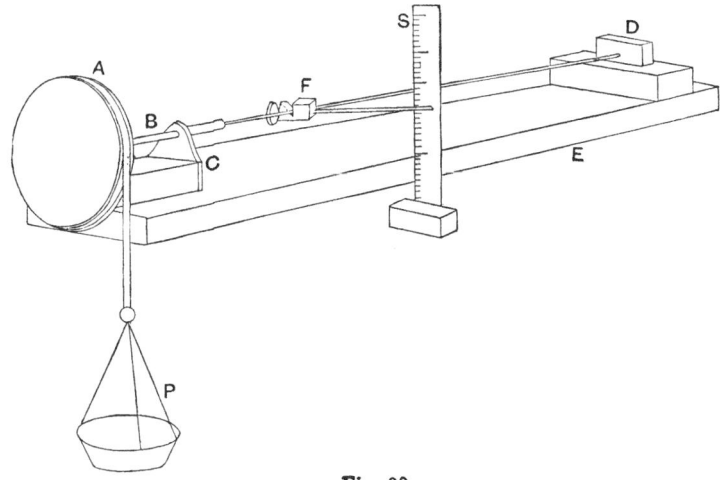

Fig. 30.

The twist of the rod is measured by means of a pointer F, which can be clamped to the rod at any point. The pointer should be adjusted so as to be approximately horizontal when half the greatest load is in the pan. A vertical millimetre scale S is used for measuring the displacement of the *tip* of the pointer*.

If the rod be not quite straight, it will bend slightly when it is twisted. Errors due to this cause may be eliminated by using a clamp fitted with *two* pointers of equal length so arranged that the line joining their tips passes approximately through the axis of the rod. The mean of the displacements of the tips of the two pointers will be free from any error due to bending of the rod.

60. Determination of rigidity. When a mass M grms. is placed in the pan, the wheel will revolve till the couple due to the

---

* The twist of the rod may also be observed by means of a mirror attached to the rod by a small clamp, the angle through which the mirror turns being observed by means of a vertical scale and a telescope with cross wires. If the distance of the scale from the mirror be d cm. and if the image of the scale move past the cross wire through z cm., when the mirror turns through ϕ radians, then $\phi = z/2d$. The results obtained with the mirror are free from any error due to bending.

elasticity of the rod balances the couple due to the mass $M + M_0$, where M_0 grms. is the mass of the pan. If the radius of the wheel be R cm., this couple is $(M + M_0) Rg$ dyne-cm., where g cm. sec.$^{-2}$ is the acceleration due to gravity.

Let the pointer be clamped at a distance of l cm. from the fixed end of the rod, and let ϕ_0 and $\phi + \phi_0$ radians be the angles through which it turns when the masses M_0 and $M + M_0$ are hung from the wheel. Then, if the rigidity of the material be n dynes per square cm. and the radius of the rod be a cm., we have, by Chapter II, § 39, equation (23),

$$\tfrac{1}{2}\pi n a^4 \phi_0 / l = M_0 g R,$$

$$\tfrac{1}{2}\pi n a^4 (\phi_0 + \phi)/l = (M + M_0) g R,$$

or, by subtraction,

$$\tfrac{1}{2}\pi n a^4 \phi / l = M g R.$$

Let the length of the pointer, measured from the axis of the rod, be p cm., and let y cm. be the vertical distance through which the tip moves, when a load M is placed in the pan. Then, if the angle between the pointer and a horizontal plane be never greater than about $\tfrac{1}{10}$ radian or 6°, we may write

$$\phi = y/p.$$

Hence, $n = \dfrac{2gpR}{\pi a^4} \cdot \dfrac{Ml}{y}$ dynes per square cm. ....(1)

The quantity $2gpR/\pi a^4$ is a constant for the given system; its value can be calculated once for all as soon as p, R and a have been measured*.

The diameter, $2R$ cm., of that part of the wheel on which the tape is wound is measured with calipers, and the mean diameter of the rod, $2a$ cm., is obtained from the readings of a screw-gauge, two perpendicular diameters being measured at several points on the rod. The proper zero correction must be applied to the mean

---

* The rod is here supposed to be truly cylindrical. When the radius is not quite constant, it will generally suffice to treat the rod as a cylinder whose radius is equal to the mean radius of the rod. The quantity we are really concerned with is not, however, the mean radius, but the fourth root of the reciprocal of the mean value of (radius)$^{-4}$. The appropriate correction is calculated in Note VI, § 2.

of the readings. The diameter of the rod must be measured carefully since the *fourth* power of the radius appears in the formula for n.

The deflexion y depends upon the two variable quantities M and l. If both M and l be varied, the observations may be combined in the following manner :—

Some value of l, say l_1, is chosen and the pointer is clamped so that the distance of the *centre* of the clamp from the nearer face of the block D is l_1 cm. The mass in the pan is then increased from zero to some maximum load by equal steps and is then diminished to zero by equal steps ; the reading of the tip of the pointer is taken at each stage. The greatest mass used should not be sufficient to give the rod any permanent twist. All these readings are *recorded* and the corresponding values of y are deduced from them by subtracting the mean of the two readings for any load from the mean of the two readings for zero load. The observations are then repeated with other lengths l_2, l_3 .... If four values of l are used, they may be approximately $\frac{1}{4}L$, $\frac{1}{2}L$, $\frac{3}{4}L$ and L, where L is the whole length of the rod. If, for a given value of l, y/M prove to be nearly constant for different loads, the mean may be taken as the best value of y/M for that value of l. When the irregularities are serious, the values of M and y for each value of l should be shown on a diagram as in Fig. 31. Since the representative points

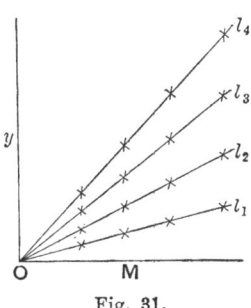

Fig. 31.

should lie, ideally, on straight lines through the origin O, a straight line is drawn by the aid of a stretched thread (§ 53) for each value of l so that the corresponding points (including the origin) are distributed as fairly as possible about it. The difference

between the ordinates of two points *on this line* corresponding to
$M = 0$ and to some definite load M (say 500 grms.) is taken as the
value of y for that load.   In this way the best value of y/M for
each value of l is found.

If the best values of y/M make lM/y nearly constant for
different values of l, the mean may be taken as the best value of
lM/y.   If the irregularities be serious, a second diagram should be
made, as in Fig. 32, showing how y/M depends upon l.   Here

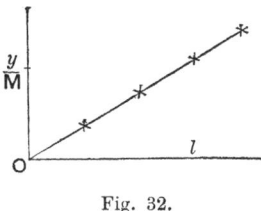

Fig. 32.

again the points should lie on a straight line through the origin,
and the best value of lM/y is found from a *straight* line drawn
in the same way as those in the first diagram.   The best value of
lM/y is used in calculating n by equation (1).

If the apparatus has been skilfully constructed, the torsion
wheel may be assumed to be truly centred on the axis.   The
effect of any small error of centering could be eliminated by
twisting the rod in *both* directions and taking the mean of the
results.

61. Practical example.   The results may be entered as in the
following record of experiments made by Messrs G. F. C. Searle and W. Burton
upon a brass rod.

Diameter of wheel $= 2R = 12\cdot0$ cm.   Hence $R = 6\cdot0$ cm.

Readings of screw-gauge for pairs of diameters of rod at right angles,

·4085	·4079	·4061	·4050	·4041	mean reading 0·4060 cm.
·4038	·4040	·4060	·4080	·4070	

Correction for zero error 0·0009 cm. to be added.

Diameter of rod $= 2a = 0\cdot4069$ cm.   Hence $a = 0\cdot2034$ cm.

Length of pointer $= p = 13\cdot92$ cm.

In the following table only the mean values of y for increasing and decreasing loads are given, but the student must *record* all the readings and deduce the displacements from them.

M grms.	$l = 10$ cm.		$l = 20$ cm.		$l = 30$ cm.		$l = 40$ cm.	
	y cm.	$\dfrac{1000y}{M}$	y cm.	$\dfrac{1000y}{M}$	y cm.	$\dfrac{1000y}{M}$	y cm.	$\dfrac{1000y}{M}$
200	0·175	·875	0·345	1·725	0·515	2·575	0·715	3·575
400	0·350	·875	0·695	1·738	1·015	2·538	1·380	3·450
600	0·520	·867	1·065	1·775	1·570	2·617	2·095	3·492
800	0·690	·862	1·410	1·762	2·060	2·575	2·800	3·500
Means	·870		1·750		2·576		3·504	

For each value of l, the table shows that y/M is practically constant, and thus the angle turned through by the pointer is proportional to the torsional couple.

Using the reciprocals of y/M, we construct a second table:

l	M/y	lM/y
10 cm.	1149 grm. cm.$^{-1}$	$1·149 \times 10^4$ grm.
20	571·4	$1·143 \times 10^4$
30	388·2	$1·165 \times 10^4$
40	285·4	$1·142 \times 10^4$

The last column shows that the angle turned through by the pointer is practically proportional to the distance between the pointer and the fixed end of the rod. The differences are probably due to want of uniformity in the rod itself and to the bending which occurs when the rod is not quite straight. As a single pointer was used, the effects of bending were not eliminated. The mean value of lM/y is $1·150 \times 10^4$ grm. Hence, by (1)

$$n = \frac{2gpR}{\pi a^4} \cdot \frac{Ml}{y} = \frac{2 \times 981 \times 13·92 \times 6·0}{\pi \times 0·2034^4} \times 1·150 \times 10^4$$
$$= 3·51 \times 10^{11} \text{ dynes per square cm.}$$

EXPERIMENT 5. **Determination of rigidity. Dynamical method.**

62. Apparatus. In this method the specimen is a wire, as straight as possible, about 50 cm. in length and about $\frac{1}{10}$ cm. in diameter. The two ends of the wire AB (Fig. 33) are soldered into two pieces of rod C, D about $\frac{1}{2}$ cm. in diameter. The rod C is held in a suitable firm clamp so that the wire is vertical, and the rod D passes through an inertia bar E, being secured by a set screw S.

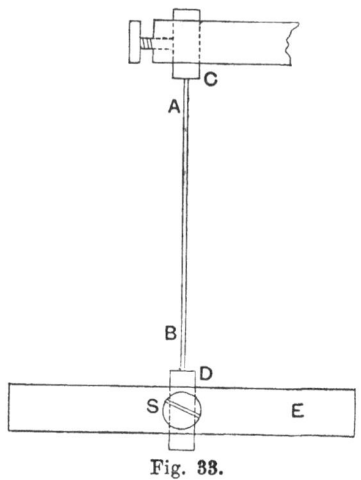

Fig. 33.

The ends of the wire itself are sometimes secured by means of set screws. If this inferior plan be adopted, care should be taken to note the exact position of each set screw relative to the corresponding end of the wire, so that the system can be taken to pieces and put together again without changing the effective length of the wire.

The compound stand shown in Fig. 34 forms a convenient support for the brass rod soldered to the upper end of the torsion wire. The stand is fitted with a moveable block which can be clamped to the upright in three different positions, and this block can be used to hold a stout rod in a vertical or a horizontal position. The small rod soldered to the torsion wire may be secured by a set screw in a hole drilled along the axis of a vertical

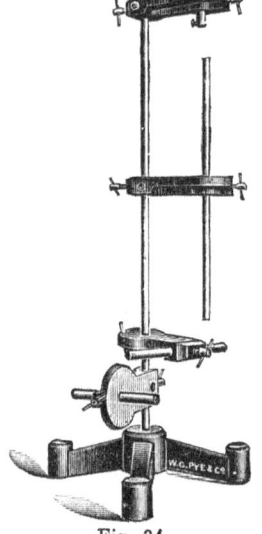

Fig. 34.

rod held in the block as shown at the top of the stand. The compound stand is very convenient for many purposes.

63. Determination of rigidity. Let the length of the wire AB, measured between the ends of the rods C, D, be l cm., let the radius of the wire be a cm. and let the moment of inertia of the inertia bar about the axis of the wire be K grm. cm.² When the bar is displaced from its equilibrium position through ϕ radians, the twist of the wire per unit length is ϕ/l radians per cm. By Chapter II, § 39, equation (23), if G be the couple which the wire exerts upon the bar,

$$G = \frac{\pi n a^4 \phi}{2l} \text{ dyne-cm.}$$

When the bar vibrates, the angular acceleration of the bar towards its equilibrium position at any time is G/K or $\pi n a^4 \phi/2lK$ radians per sec. per sec. (Note III, § 2), and thus the angular acceleration is proportional to the angular displacement. Hence the motion is harmonic, and by Note V, § 2, the time of a complete vibration is given by

$$T = \frac{2\pi}{\sqrt{\text{angular acceleration for one radian}}}$$

$$= 2\pi \sqrt{\frac{2lK}{\pi n a^4}}.$$

Deducing the rigidity, n, from this equation, we have

$$n = \frac{8\pi K l}{T^2 a^4}. \quad\ldots\ldots\ldots\ldots\ldots\ldots\ldots(1)$$

The length of the wire, l cm., is found with a centimetre scale, and its mean diameter, $2a$ cm., is obtained from the readings of a screw-gauge, two perpendicular diameters being measured at several points on the wire. The proper zero correction must be applied to the mean of the readings. The diameter of the wire must be measured carefully since the fourth power of the radius appears in formula (1)*.

The moment of inertia of the inertia bar is calculated from its

* See Footnote on page 92.

mass and its dimensions by the methods of Note IV. If the bar be rectangular, and if its length be $2L$ cm., its width $2A$ cm. and its mass M grms., and if K be its moment of inertia about an axis through its centre at right angles to $2L$ and to $2A$,

$$K = \tfrac{1}{3} M \left(L^2 + A^2\right) \text{ grm. cm.}^2.$$

If the bar be a solid circular cylinder, of length $2L$ and radius R, with its axis at right angles to the axis of the wire,

$$K = M \left(\tfrac{1}{3} L^2 + \tfrac{1}{4} R^2\right) \text{ grm. cm.}^2.$$

For rough purposes it is sufficient to take M as the mass of the system which is detached from the torsion wire when the set screw is slackened. For more accurate work, M should be the mass of the bar before any holes are bored in it. (See Note VII.)

If the length of the inertia bar be large compared with its width or its diameter, the term $\tfrac{1}{3} M L^2$ is the chief term in the expression for K, and the terms involving A^2 or R^2 are comparatively very small. Hence it is quite unnecessary to measure A or R with a screw-gauge; it is sufficient to use a millimetre scale. On the other hand, $2L$ should be measured as accurately as possible.

The time occupied by a large number of *complete* vibrations is found at least twice, the observation in each case extending over at least three minutes. The mean time of a complete vibration (T sec.) is then deduced. Unless the time-piece used be known to be keeping good time, it should be compared with a good clock to find the necessary correction.

A stop-watch is generally used in observing the time of vibration. But very good results can be obtained by the following method with an ordinary watch or clock fitted with a seconds hand. At every fifth transit, from left to right, of one end of the bar past a fixed mark, the time indicated by the watch is observed. After a sufficient number of these times have been recorded, the time of the 0th transit is subtracted from that of the 50th and the time of the 5th transit is subtracted from that of the 55th, and so on. In this way we obtain a number of intervals, each corresponding to 50 complete vibrations. With careful work these intervals will agree closely and their mean will furnish a reliable value of the time of 50 complete vibrations. The following experimental results will illustrate the working of the method.

Transit	Time		Transit	Time		Interval	
	min.	sec.		min.	sec.	min.	sec.
0	1	42·2	50	4	37·2	2	55·0
5	2	0·0	55	4	54·7	2	54·7
10	2	17·2	60	5	12·3	2	55·1
15	2	35·0	65	5	29·8	2	54·8
20	2	53·0	70	5	48·0	2	55·0
25	3	10·0	75	6	4·0	2	54·0
30	3	27·7	80	6	22·2	2	54·5
35	3	45·0	85	6	39·5	2	54·5
40	4	2·4	90	6	57·0	2	54·6
45	4	20·0	95	7	14·3	2	54·3

Mean 2 min. 54·65 sec.

The mean time of 50 complete vibrations is 174·65 seconds and hence the periodic time is 3·493 seconds. In this example 10 *independent* observations of the time of 50 vibrations have been made in the time occupied by 95 vibrations. This method has the advantage that it involves no interference with the regular working of the watch such as occurs when a stop-watch is started or stopped.

64. Practical example. The observations may be entered as in the following record of an experiment on a brass wire.

Length of wire under torsion $= l = 52·2$ cm.

Readings of screw-gauge on wire, for pairs of diameters at right angles,

| 0·1218 | 0·1219 | 0·1222 |
| 0·1219 | 0·1220 | 0·1222 |

. Mean reading 0·1220 cm.

Correction for zero error, 0·0003 cm., to be added.

Hence diameter $= 2a = 0·1223$ cm. Radius $= a = 0·06115$ cm.

Mass of cylindrical bar $= M = 649·1$ grms.

Length $= 2L = 37·82$ cm. Hence $L = 18·91$ cm.

Diameter $= 2R = 1·60$ cm. Hence $R = 0·80$ cm.

Hence $K = M\left(\frac{1}{3}L^2 + \frac{1}{4}R^2\right) = 649·1\,(119·2 + 0·2) = 7·75 \times 10^4$ grm. cm.^{2}.

Time of 40 complete vibrations 183·0, 182·4, 182·4. Mean 182·6 sec.

Hence $T = 4·565$ sec.

Thus, by (1)

$$\text{Rigidity} = n = \frac{8\pi Kl}{T^2 a^4} = \frac{8\pi \times 7·75 \times 10^4 \times 52·2}{4·565^2 \times 0·06115^4} = 3·49 \times 10^{11} \text{ dynes per square cm.}$$

The value of the diameter of the wire is subject to an uncertainty of about 2 parts in 1000. Due to this cause, there is an uncertainty in the value of n of about 8 parts in 1000, since n is inversely proportional to the fourth power of the radius. The number 3·49 is therefore uncertain to the amount 0·03.

EXPERIMENT 6. **Determination of Young's modulus by uniform bending of a rod. Statical method.**

65. Apparatus. In order to produce uniform bending in a rod it is necessary that the "bending moment" (Chapter II, § 31) should have a constant value at every point of the rod. This condition is easily secured if the rod be bent in the manner indicated in Fig. 35. The rod AB rests symmetrically on two knife edges C, D, which are fixed to a stout bed XY.

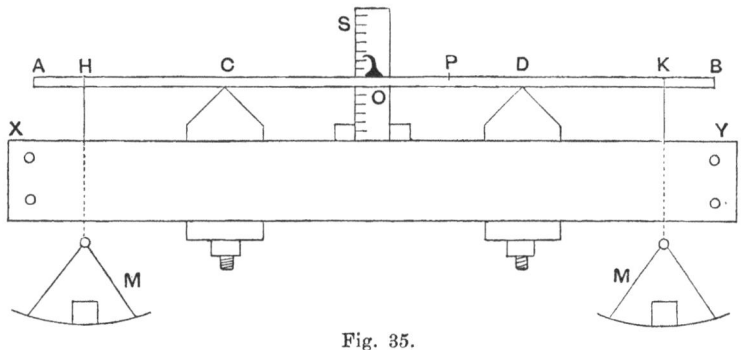

Fig. 35.

The most suitable bed for this and for other experiments on the bending of rods is a small lathe bed, but a good substitute may be constructed of two wooden beams about 120 cm. long, 5 cm. wide, and 15 cm. deep. The beams are bolted together at each end, a piece of wood about 2 cm. thick being placed between them at each end so that there is a gap of 2 cm. between the beams to allow of the passage of vertical strings. The knife edges may be short pieces of angle iron firmly screwed to two boards which are secured to the bed by bolts passing through the gap in the bed, as shown in the figure. The bed rests on two blocks of a convenient height.

The rod is bent by means of two equal masses placed in light scale-pans suspended from the two points H, K on the rod, the distance HC being equal to KD. To determine the vertical displacement of the middle point O, a pin is fixed by wax to the rod at O, the pin being bent so that the part near the tip is horizontal, and a vertical scale S is set up near O in such a position that the tip of the pin is close to the scale.

Errors of parallax may be avoided and the accuracy of the readings may be increased by taking the scale readings of the pin by means of a fixed telescope. The distance of the telescope from the scale should be as small as the focussing of the telescope will allow, in order that the magnification may be as great as possible.

When a circular rod is used, a cross-bar about 4 cm. long and 1 cm. wide should be soldered or otherwise fixed to the rod. If the cross-bar rest on one of the knife edges, it prevents the rod from rolling.

A scale holder convenient for many purposes is shown in Fig. 36. It consists of a rectangular block of brass about 5 cm.

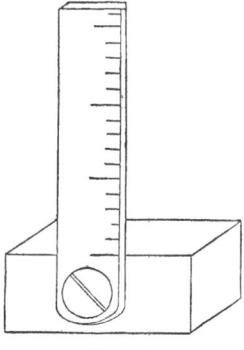

Fig. 36.

in length, and 2·5 cm. in width and depth. A steel scale, divided in millimetres*, is secured to the block by a screw passing through a hole at one end of the scale. The scale can be used in a number of different positions. The screw may conveniently have a milled head so that it can be tightened by hand.

---

* Steel scales divided to *half* millimetres should be avoided, unless the dividing lines are very fine.

66. Determination of Young's modulus. Let the mass
of each pan be M_0 grammes and let the mass in *each* pan be
M grammes. Let each of the distances HC and KD (Fig. 35),
when measured horizontally, be p cm. Since the system is
symmetrical about O, the middle point of the rod, the force on
each knife edge is the same and thus the part of this force which
is due to the pans and the loads is $(M_0 + M)g$ dynes.

If P be any point of the rod between C and D, and if G be the
part of the bending moment at P which is due to the pans and
the loads,

$$G = (M_0 + M)(PK - PD)g = (M_0 + M)pg \text{ dyne-cm. } ...(1)$$

Hence the bending moment due to the pans and the loads has the
same value at all points of the rod between C and D.

In addition to G, there is the bending moment due to the weight
of the part of the rod between P and B acting at a point midway between P and B, as well as the bending moment due to a
vertical force at D equal to half the weight of the rod. When
the effects due to these moments are small, it follows from Hooke's
law, Chapter I, § 2, that, at each point of the rod, any small change
of curvature of the axis of the rod due to the pans and the loads is
the same as if the rod were without weight. The curvature is
measured by $1/\rho$, the reciprocal of the radius of curvature of the
axis, and is zero when the axis is straight. Since G is constant at
all points between C and D, this change of curvature is constant.
Hence, when the rod does not bend appreciably under its own
weight, we may treat it as if it were weightless, when we discuss
the effects of small loads applied to it. In what follows, we shall
neglect the weight of the rod.

If the "moment of inertia" of the transverse section of the
rod, about an axis passing through the centre of gravity of that
section and at right angles to the plane of bending, be I cm.[4], we
have, by Chapter II, § 31,

$$G = \frac{EI}{\rho} \text{ dyne-cm., } \quad ....(2)$$

where ρ cm. is the radius of curvature of the neutral filament of
the rod. Inserting the value of G given by (1), we have

$$E = \frac{G\rho}{I} = \frac{(M_0 + M)gp\rho}{I} \text{ dynes per square cm. } ......(3)$$

The curvature $1/\rho$ is easily deduced from the vertical displacement of the middle point O due to the two loads. Suppose that the point of the pin fixed to the rod at O moves through the vertical distance h_0 cm., when the pans alone are hung from the rod, and that it rises through a further distance h when a mass of M grammes is placed in each pan. Then, if the distance CD between the knife edges be $2l$ cm., we have, by the geometry of the circle,

$$(h_0 + h)(2\rho - h_0 - h) = l^2.$$

In most cases $h_0 + h$ will be negligible in comparison with 2ρ and then we may write

$$\rho = \frac{l^2}{2(h_0 + h)} . \qquad\qquad (4)$$

From (3) and (4) we find

$$E = \frac{(M_0 + M)\, pl^2 g}{2I(h_0 + h)} \text{ dynes per square cm.} \qquad (5)$$

Since, by (5), the elevation is proportional to the load,

$$\frac{M_0 + M}{h_0 + h} = \frac{M_0}{h_0} = \frac{M}{h},$$

and thus $\qquad E = \dfrac{Mpl^2 g}{2Ih}$ dynes per square cm. *............(6)

When the rod is of circular section, with diameter $2a$ cm., the "moment of inertia," I, of the area of the section about the axis in the plane of the section which passes the centre, is given by †

$$I = \tfrac{1}{4}\pi a^4 \text{ cm.}^4 . \qquad\qquad (7)$$

Hence, by (6), for a circular rod

$$E = \frac{2Mpl^2 g}{\pi a^4 h} \text{ dynes per square cm. } \qquad (8)$$

When the rod is of rectangular section with sides $2a$ and $2b$ cm., the side $2b$ being vertical when the rod is in position for bending, the "moment of inertia" of the area about an axis through its centre parallel to the side $2a$ is given by

$$I = \tfrac{2}{3}ab^3 \text{ cm.}^4 . \qquad\qquad (9)$$

* See Note XI.　　　　† See Note IV, § 12.

Hence, by (6) for a rectangular rod

$$E = \frac{3Mpl^2g}{8ab^3h} \text{ dynes per square cm. } \dots\dots\dots\dots(10)$$

A series of observations for h is made. The masses in the pans are increased by four or five equal steps from zero to some maximum value which does not strain the rod beyond the elastic limit, and the masses are then diminished to zero by the same steps, a reading of the pin being taken at each stage. The masses should be placed gently in the pans to avoid the extra stresses which occur when the masses are dropped into the pans[*]. The difference between the mean of the two readings for given masses and the mean of the two readings, when the pans are empty, is taken as the elevation due to these masses.

Care must be taken not to load one pan so much more than the other that the greater overbalances the smaller load.

If h/M prove to be nearly constant for different loads, the mean value may be taken as the best value of h/M to use in calculating E. When there are serious irregularities, the values of M and of h should be shown on squared paper, and a straight line should be drawn by aid of a stretched thread, as in § 53, so as to pass as evenly as possible among the plotted points. The difference between the values of h *as shown by this line* for $M = 0$ and for some definite mass M is taken as the value of h for that mass. These values of M and h are used in (8) or (10).

If we are to keep within the elastic limit, the maximum elongation of the most highly strained longitudinal filaments should not exceed about $\frac{1}{1000}$ cm. per cm. It follows, by Chapter II, § 30, that ρ must not be less than $1000d$, where $2d$ stands either for the diameter of a circular rod or for the vertical thickness of a rectangular one. Hence, by (4), we see that $h_0 + h$ should not exceed $l^2/(2000d)$. Thus, if $2d = 1$ cm. and if $2l = 80$ cm., $h_0 + h$ should not exceed 1·6 cm.

67. Mirror method of determining curvature. We have just seen that the elevation of the middle point of the rod must be comparatively small, if the strains are not to pass the elastic

---

[*] See the first Footnote on page 82.

limit, and thus, if a millimetre scale, read by eye, is the only available means of measuring the elevation, it is clear that no great accuracy is possible. The distance to be measured is, however, easily increased by using a mirror. A plane mirror R (Fig. 37)

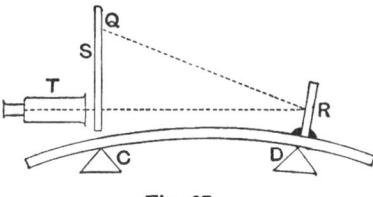

Fig. 37.

is attached to the rod with wax immediately over the knife edge D and is adjusted so that the normal to the mirror is approximately parallel to the rod when the rod is not loaded. A vertical scale S is placed over the other knife edge C and a telescope T, fitted with cross wires, is placed so that the observer can view a point Q on the scale by reflexion at the mirror. In this method we rely on the axis of the telescope remaining in a fixed direction. Care should, therefore, be taken that the telescope is *firmly* mounted. It will be difficult to obtain satisfactory readings if the apparatus be much disturbed by vibration.

Since the tangent to the rod at the middle point remains horizontal and since $CD = 2l$, it follows that, if the tangent at D turn through an angle θ when the load $M_0 + M$ is hung at each end,

$$\theta = l/\rho.$$

But, if the scale appear to move past the cross wire of the telescope through z_0 cm., when the scale-pans are hung on, and through an additional z cm., when a mass M is placed in each pan, we have, for small angles,

$$\frac{z_0 + z}{2l} = 2\theta,$$

since the angle turned through by RQ is twice the angle turned through by the mirror. Hence,

$$\rho = \frac{l}{\theta} = \frac{4l^2}{z_0 + z}.$$

Comparing this result with (4), we have

$$z_0 + z = 8 (h_0 + h),$$

or, since this holds for all corresponding values of h and z,

$$z = 8h.$$

Thus the apparent movement of the scale past the cross wire is eight times the corresponding motion of the middle point of the rod.

Substituting for h in (8) and (10) we have for a round rod

$$E = \frac{16Mpl^2g}{\pi a^4 z} \text{ dynes per square cm. } \quad\text{.........(11)}$$

and for a rectangular rod

$$E = \frac{3Mpl^2g}{ab^3 z} \text{ dynes per square cm. } \quad\text{............(12)}$$

The readings obtained by aid of the mirror may be treated as those obtained by aid of the pin are treated in § 66.

68. Practical example. The observations may be entered as in the following record of an experiment made by Mr D. L. H. Baynes upon a circular rod of steel.

Readings of screw-gauge for pairs of diameters at right angles,

·9562	·9612	·9585	·9557	·9556
·9620	·9595	·9651	·9568	·9596

. Mean reading ·9590 cm.

Correction for zero error 0·0006 cm.; to be added.
Mean diameter $= 2a = ·9596$ cm.   Radius $= a = ·4798$ cm.
Distance between knife edges $= 2l = 80$ cm.
Distance between knife edge and point of suspension of load $= p = 35$ cm.

Load M grms.	Scale reading Loading cm.	Scale reading Unloading cm.	Mean elevation h cm.	$\dfrac{h}{M}$ cm. grm.$^{-1}$
0	5·50	5·50	0·00	
500	5·32	5·31	0·18	$3·60 \times 10^{-4}$
1000	5·18	5·18	0·32	3·20
1500	5·00	5·00	0·50	3·33
2000	4·84	4·82	0·67	3·35
2500	4·69	4·69	0·81	3·24
3000	4·52	4·50	0·99	3·30
3500	4·37	4·35	1·14	3·26
4000	4·21	—	1·29	3·22

When the mean elevation h was plotted against the load M, the straight line lying most evenly among the points cut the line $M=0$ at 0·02 cm. and the line $M=4000$ at 1·30 cm., the difference being 1·28 cm. Hence, for calculation, we use $h/M=1\cdot28/4000=3\cdot20\times10^{-4}$ cm. grm.$^{-1}$. The values of h/M given in the table are somewhat irregular. If the value $3\cdot60\times10^{-4}$ be excluded, the mean of the remainder is $3\cdot27\times10^{-4}$, slightly higher than the number obtained from the diagram. Using $h/M=3\cdot20\times10^{-4}$, we find, by (8), for Young's modulus

$$E=\frac{2Mpl^2g}{\pi a^4 h}=\frac{2\times35\times40^2\times981}{3\cdot142\times0\cdot4798^4\times3\cdot20\times10^{-4}}=2\cdot06\times10^{12}\text{ dynes per square cm.}$$

EXPERIMENT 7. **Determination of Young's modulus by uniform bending of a rod. Dynamical method.**

69. Determination of Young's modulus. The ends of the wire or rod are soldered into two clamping-screws which are secured to two equal inertia bars AB, CD (Fig. 38). Two light

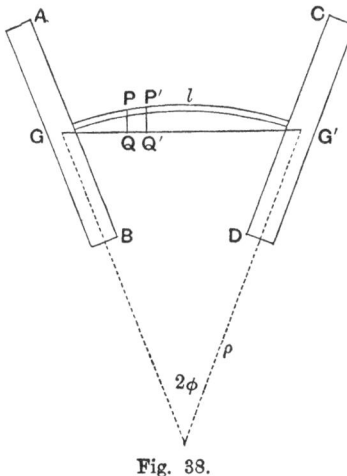

Fig. 38.

hooks about 4 cm. long are screwed into the bars at G, G', so that the hooks are perpendicular to the wire, as in Fig. 39, which shows a section of the arrangement by a plane through G perpendicular to the axis of the bar AB. The cylindrical recess in the inertia bar allows the end of the clamping-screw to lie on the axis of the bar. By means of the hooks, the system is suspended by two parallel strings at least 50 cm. long. Since the centres of gravity of the bars are below the hooks, the system can rest in stable equilibrium with the plane $ABCD$ horizontal.

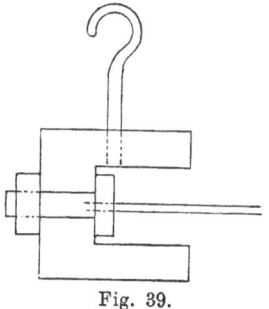

Fig. 39.

If the two bars be now turned through equal angles ϕ in opposite directions and be then set free, the system will vibrate, each bar executing harmonic vibrations in a horizontal plane.

When the vibrations are small, the rod or wire is only slightly bent, and hence the distance GG', measured along the straight line, differs very little from the length of the wire itself. If P, P' be two neighbouring points on the wire and Q, Q' be their projections on the straight line GG', we have $QQ' = PP' \cos \psi$, where ψ is the angle between GG' and the tangent at P. Thus, since ψ is small,

$$PP' - QQ' = PP' (1 - \cos \psi) = \tfrac{1}{2} PP' \cdot \psi^2.$$

Now, the maximum value of ψ occurs at the ends of the wire, and there $\psi = \phi$. Hence, if the length of the wire be l cm.,

$$l - GG' < \tfrac{1}{2} l \phi^2,$$

so that, when ϕ is of the first order of small quantities, the displacements of G and G' towards each other are of the second order. For small vibrations we may, therefore, treat the distance GG' as invariable.

When the mass of the wire is negligible compared with that of the bars, the motion of G and G' at right angles to GG' may be neglected.

Since the horizontal displacements of G and G' are very small compared with the length of the supporting strings, the vertical motion of G and G' is negligible.

Now, whatever be the forces acting on either bar, they may, by Note I, be reduced to a force and a couple. By Note III, § 1, the force is Mf, where M is the mass of the bar and f is the acceleration of the centre of gravity of the bar. But, to our order of accuracy, the centres of gravity of the bars are at rest, and hence the action of the wire on either bar is simply a couple which, by symmetry, must have a vertical axis. Conversely, each bar exerts a couple on the wire. Hence, the "bending moment" (Chapter II, § 31) is the same at every point of the wire, and thus the neutral filament of the wire is bent into a circular arc.

It is shown in Chapter II, § 31, that the bending moment is EI/ρ, where ρ cm. is the radius of the arc, E is Young's modulus, and I cm.^{4} is the "moment of inertia" of the area of cross section of the wire about an axis through the "centre of gravity" of that area perpendicular to the plane of bending. This axis is perpendicular to the plane of Fig. 38. If ϕ be the angle turned through by either bar from its equilibrium position, we see from Fig. 38 that $\rho = l/2\phi$, since the length of the wire is l cm.

Let the moment of inertia of either bar about a vertical axis through its centre of gravity be K gramme cm.^{2} and let the angular acceleration of the bar towards its equilibrium position be a radian sec.$^{-2}$ when the displacement is ϕ radians. Then, since the couple on the bar is EI/ρ, we have, by Note III, § 2,

$$a = \frac{\text{couple}}{\text{moment of inertia}} = \frac{EI}{K\rho} = \frac{2EI}{Kl}\,\phi.$$

The angular acceleration per radian of displacement is $2EI/Kl$ radian sec.$^{-2}$ towards the equilibrium position, and thus, if T_1 be the periodic time, we have, by Note V, § 2,

$$T_1 = 2\pi\,(\text{angular acceleration for one radian})^{-\frac{1}{2}}$$

$$= 2\pi\,\sqrt{\frac{Kl}{2EI}}. \quad\dots\dots\dots\dots\dots\dots\dots\dots\dots\dots\dots(1)$$

If the radius of the circular section of the wire be a cm., we have, by Note IV, § 12,

$$I = \tfrac{1}{4}\pi a^4 \text{ cm.}^4.$$

Hence, by (1), $E = \dfrac{8\pi Kl}{T_1^2 a^4}$ dynes per square cm. $\dots\dots\dots\dots(2)$

In (2) the mass of the wire has been entirely neglected. It may be shown* that, when m, the mass of the wire, is small compared with the mass of either bar,

$$E = \frac{8\pi l}{T_1^2 a^4} \left(K + \frac{ml^2}{60} \right). \quad \dots\dots\dots\dots\dots(3)$$

70. Experimental details. In order that the most highly strained portions of the wire should not be strained beyond the elastic limit, it is necessary that the amplitude of the vibrations of the bars should be small. As it is impossible to observe the time of vibration of the bars satisfactorily unless the motion we are considering is undisturbed by any other motion of the system, the vibration must be started without giving the bars any motion of translation. The two ends B, D (Fig. 38) are drawn slightly together by a loop of cotton thread, and the system, thus constrained, is brought carefully to rest. The desired vibration is then started by burning the thread. A pointer should be set up close to the end of one of the bars, and the transits past the pointer of a mark on the bar should be observed in finding the time of vibration. Some care is necessary in this part of the work, for it is found that with large arcs of vibration the periodic time is appreciably greater than for small arcs. If the arcs are large the result may be considerably in error.

The relation between the amplitude of the vibration and the maximum elongation of the material of the wire is easily found. For, by Chapter II, § 30 and equation (1) of § 29, if e be the maximum elongation, $e = a/\rho$ cm. per cm. But $\rho = l/2\phi$ and hence $\phi = el/2a$. Errors will be avoided if e never exceeds $\frac{1}{5000}$. If the wire be 25 cm. long and 0·1 cm. in diameter, e will not exceed $\frac{1}{5000}$, if ϕ does not exceed $\frac{1}{20}$ radian or about 3°.

The time of vibration may be found by a stop-watch or in other ways (see § 63), but unless the time-piece is known to be keeping correct time, it should be compared with a standard clock.

Readings for the diameter of the wire should be made by a screw-gauge at four or five places equally spaced along the wire, two perpendicular diameters being measured at each place. The mean reading, when corrected for the zero error of the gauge, is taken as the diameter, $2a$, of the wire*.

If the readings show that the diameter of the wire is sensibly elliptical, a mark should be made on one of the clamping-screws, and the periodic time should be observed when the clamping-screws are adjusted so that the mark is vertical. The screws should then be loosened and the wire turned about its axis until the mark is horizontal, the screws being then tightened and the periodic time again observed. The mean of the two periodic times is used in (2).

The mass of each bar is found before the hole is bored in it and before the hook is fixed to it, and the value of this mass is stamped on the bar†. If the mass of the bar be M grms., if its length be $2L$ cm and if the sides of the square section be $2A$ cm., we have, by Note IV, § 6,

$$K = \tfrac{1}{3} M (L^2 + A^2) \text{ gramme cm.}^2.$$

The simple theory supposes that K_A and K_C, the moments of inertia of the bars AB, CD (Fig. 38), are exactly equal. This will not generally be the case in practice, but it follows from the principles employed in obtaining equation (3) that, when K_A and K_C are nearly equal, the observed time of vibration will not differ appreciably from that which would be found if the moment of inertia of each bar were $\tfrac{1}{2} (K_A + K_C)$. We may therefore take K in equation (2) as equal to the mean of the moments of inertia of the two bars.

71. Practical example.   The observations may be entered as in the following record of an experiment made by Mr D. L. H. Baynes on a wire of German-silver.

Length of bars $= 2L = 32\cdot10$ cm.   Breadth of bars $= 2A = 1\cdot29$ cm.

Mass of each bar $= M = 441$ grammes.

Moment of inertia of each bar $= K = \tfrac{1}{3} M (L^2 + A^2) = \tfrac{1}{3} 441 (16\cdot05^2 + \cdot64^2)$
$\qquad\qquad\qquad\qquad = 3\cdot793 \times 10^4$ grm. cm.^{2}.

Length of wire $= l = 31\cdot15$ cm.

Readings of screw-gauge for pairs of diameters at right angles,

·1189	·1187	·1187	·1188	mean ·1188 cm.	mean ·1189 cm.
·1189	·1190	·1191	·1189	mean ·1190 cm.	

---

* See Footnote on page 92.

† The reasons for this procedure are similar to those explained in Note VII.

Correction for zero error ·0005 cm.; to be added.

Mean diameter $=2a=·1189+·0005=·1194$ cm.   Radius $=a=·0597$ cm.

The mean readings for the two diameters, viz. ·1188 and ·1190 cm. were so nearly equal that it was considered unnecessary to change the positions of the clamping-screws in the bars.

Time of 50 complete vibrations 71·0, 70·8.   Mean 70·9 secs.

Periodic time $T_1=1·418$ secs.

Hence, by (2), we find for Young's modulus

$$E=\frac{8\pi Kl}{T_1^2 a^4}=\frac{8\times3·142\times3·793\times10^4\times31·15}{1·418^2\times0·0597^4}=1·16\times10^{12} \text{ dynes per square cm.}$$

EXPERIMENT 8. **Comparison of elastic constants. Dynamical method.**

72. Method. By the apparatus used for EXPERIMENT 7 we can compare Young's modulus E and the rigidity n of the material of the wire by simply observing two times of vibration. The inertia bars are unhooked from the strings and one bar is clamped to a shelf or other suitable support so that the wire is vertical. The other bar is then caused to vibrate about a vertical axis, exactly as in EXPERIMENT 5 for finding the rigidity. Since the vibrating bar is of square section, its moment of inertia about the wire is equal to K, its moment of inertia about the axis of the hook, provided that the effects of the hook, the clamping screw, and the recess, be negligible. If T_2 be the periodic time of the torsional vibrations, we have, by equation (1), § 63,

$$n=\frac{8\pi Kl}{T_2^2 a^4}. \qquad\qquad\qquad\qquad\dots\dots\dots\dots\dots(1)$$

The periodic time, T_1, of the vibrations discussed in § 69 is then observed. By equation (2), § 69,

$$E=\frac{8\pi Kl}{T_1^2 a^4}, \qquad\qquad\qquad\qquad\dots\dots\dots\dots\dots(2)$$

and hence, by (1), $\dfrac{E}{n}=\dfrac{T_2^2}{T_1^2}. \qquad\qquad\qquad\dots\dots\dots\dots\dots(3)$

Thus we can compare E and n without knowing the length or diameter of the wire or the dimensions of the inertia bars.

If K_A and K_C, the moments of inertia of the two bars, be not exactly equal, each bar should be caused to vibrate in turn. The mean of the two periodic times will be very nearly equal to that which would be found if the moment of inertia of each bar were $\frac{1}{2}(K_A + K_C)$. We may therefore take T_2 in equation (1) as equal to the mean of the two periodic times. On page 111 it is shown that in equation (2) we may take K as equal to $\frac{1}{2}(K_A + K_C)$.

When E/n has been found, Poisson's ratio, σ, is easily calculated by formula (11) of § 19, Chapter I, viz.

$$\sigma = \frac{E}{2n} - 1. \quad\dots\dots\dots\dots\dots\dots\dots(4)$$

The following table* gives some values of E/n obtained by this method when applied to wires about 0·1 cm. in diameter. To make the results more complete, the values of E and n were calculated by (2) and (1), the unit for each modulus being one dyne per square cm. In each case the value of E has been corrected for the mass of the wire, the correction being about $\frac{1}{10}$ per cent.

Material	E dyne cm.$^{-2}$	n dyne cm.$^{-2}$	E/n	σ
"Silver"-steel	$1\cdot981 \times 10^{12}$	$7\cdot872 \times 10^{11}$	2·516	0·258
Brass (hard-drawn)	1·022	3·715	2·751	0·376
Phosphor-bronze	1·201	4·359	2·755	0·378
Silver (hard-drawn)	0·778	2·816	2·762	0·381
Copper (hardened by stretching)	1·240	3·880	3·195	0·598
Copper (annealed)	1·292	4·018	3·217	0·608
Nickel (hardened by stretching)	2·395	7·424	3·227	0·614
Platinoid...	1·359	3·602	3·773	0·887
German-silver (hard-drawn) ...	1·155	2·618	4·414	1·207

For the last five substances E/n is greater than 3, and hence, by (4), σ is greater than $\frac{1}{2}$. For an isotropic material we have, by formula (12) of § 19, Chapter I,

$$3k(1 - 2\sigma) = 2n(1 + \sigma).$$

Thus, if σ were greater than $\frac{1}{2}$, either the bulk modulus k or the rigidity n would be negative. In the first case, a hydrostatic pressure applied to the material would cause it to increase in volume, and in the second case, a positive shearing stress would give rise to a negative shear. We infer that the wires of the last five substances are so far from being isotropic that the theory of isotropic

* G. F. C. Searle, *Philosophical Magazine*, Feb. 1900, p. 199.

solids does not furnish a working approximation. Considering how violent is the process of wire-drawing, this result is hardly surprising.

73. Practical example. The results may be entered as in the following record of an experiment by Mr D. L. H. Baynes on the wire of German-silver used in the experiment of § 71.

Time of 50 complete vibrations (Young's modulus) 71·0, 70·8. Mean 70·9 secs.

Hence $T_1 = 1·418$ secs.

Time of 50 complete vibrations (Rigidity) 153·4, 153·4. Mean 153·4 secs.

Hence $T_2 = 3·068$ secs.

By (3)
$$\frac{E}{n} = \frac{T_2^2}{T_1^2} = \frac{3·068^2}{1·418^2} = 4·68.$$

Hence, by (4), Poisson's ratio $= \sigma = (E/2n) - 1 = 1·34$.

Since the maximum value of σ for an isotropic solid is $\frac{1}{2}$, this result shows that the German-silver wire is far from being isotropic.

EXPERIMENT 9.   **Determination of Poisson's ratio by the bending of a rectangular rod.**

74. Introduction. It is shown in Chapter II, § 29, that, when a rod of rectangular section is bent into a circular arc, the transverse section is distorted*. The sides BC, AD (Fig. 12) of the section, which are initially parallel to the axis of bending, become circular arcs having a common centre S, while the sides AB, CD which are initially perpendicular to the axis of bending become straight lines $A'B', C'D'$ passing through S. If the distance of S from O, the point where the neutral filament cuts the plane of the diagram, be ρ' cm. and if the radius of curvature of the neutral filament be ρ cm., then we have, by formula (5) of § 29,

$$\sigma = \frac{\rho}{\rho'}. \quad \ldots\ldots\ldots\ldots\ldots\ldots\ldots(1)$$

Hence we can find Poisson's ratio σ, if we measure the longitudinal curvature $1/\rho$, viz. the curvature of the neutral filament, and determine the point S through which the sides $A'B', C'D'$ would pass, if continued. Instead of finding S, we may deduce $1/\rho'$ from

---

* See Chapter II, §§ 33, 34, for the difference between a rod and a blade with respect to the distortion of the section.

the angle θ between the sides $A'B'$ and $C'D'$ when the rod is bent. For, if the width (BC) of the bar be $2a$ cm., we have

$$\frac{1}{\rho'} = \frac{\theta}{2a} \text{ cm.}^{-1}. \quad \dots\dots\dots\dots\dots\dots\dots(2)$$

We shall call $1/\rho'$ the transverse curvature.

The apparatus is arranged as in Fig. 40. The rod rests upon two knife edges N, N' and can be bent by means of two equal masses placed in the pans which hang from the stirrups L, L' near the ends of the rod, the distances LN and $L'N'$ being equal. The rod should be 0·2 to 0·3 cm. in thickness and 2 to 4 cm. in width. The general

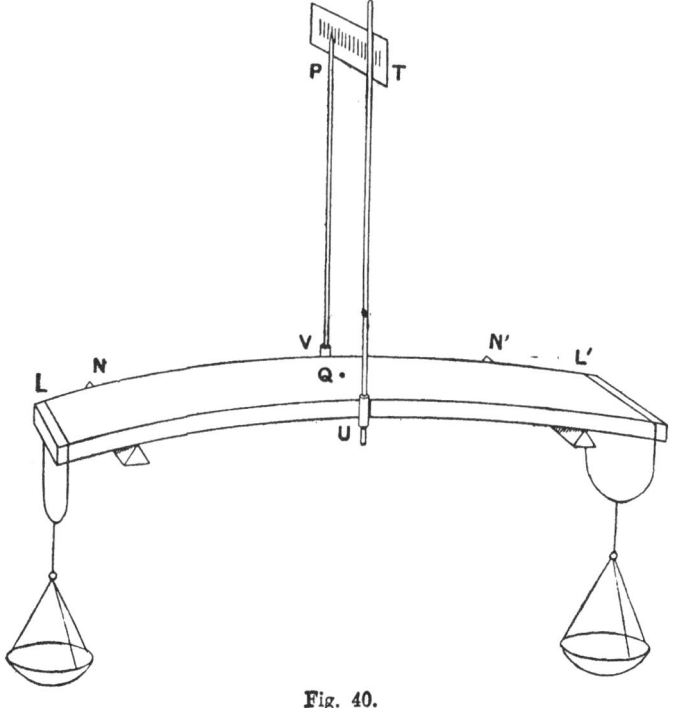

Fig. 40.

arrangement of the apparatus is the same as that in EXPERIMENT 6, to which the reader should refer. At U, V, midway between the knife edges, are fixed two steel needles about 3 mm. in diameter and 40 to 50 cm. in length. The needle fixed at U

8—2

carries a horizontal scale T and the point P of the other needle moves along this scale when the rod is bent, the relative motion indicating the extent of the distortion of the transverse section. The tip of the needle VP should be ground to a fine point so that the readings may be taken with certainty to $\frac{1}{10}$ mm.

An efficient method of attaching the needles to the rod is shown in Fig. 41. Two connectors, each fitted with two set screws, such

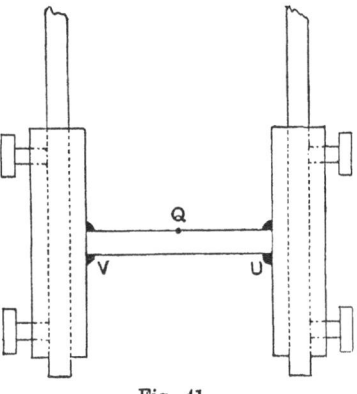

Fig. 41.

as are used for electrical connexions, are soldered to the sides of the rod at U and V and the needles are secured in the connectors by the set screws.

It would be difficult, without some special device, to solder the connectors to the rod one at a time, for the act of soldering the second connector to the rod would probably cause the solder holding the first one to melt. To avoid this trouble, the connectors may be fixed to the two needles as in Fig. 41, and the needles may then be secured in a suitable clamp so as to hold the connectors in position against the sides of the rod. The soldering may then be accomplished by aid of a soldering bit or blowpipe.

Since the motion of the pointer VP along the scale T is only small, a telescope should be used to magnify the scale and to avoid errors of parallax.

The two long needles are very sensitive to vibration, and thus it is impossible to obtain accurate readings if the apparatus be set up in a part of the laboratory which is subject to much vibration.

Since the distances LN, $L'N'$ (Fig. 40) are equal, and since the masses suspended from L and L' are equal, it follows, as in § 66, that the bending moment is constant for all points of the rod between N and N'. Hence the neutral filament is bent into a circular arc.

The value of $1/\rho'$, the transverse curvature, is deduced from the motion of the pointer VP relative to the scale T. On VS, US (Fig. 42) take P, P' such that $VP = UP' = p$, where p cm. is the

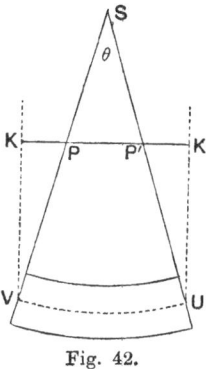

Fig. 42.

length of the steel pointer measured from the tip to the centre of the edge of the rod, and let the straight line PP' cut the vertical lines through V and U in K and K'. Then, since the angle VSU or θ is the sum of the very small angles PVK, $P'UK'$, we may take

$$p\theta = PK + P'K'.$$

If the displacement of the tip of the pointer along the scale be x cm., we may write

$$x = PK + P'K' = p\theta.$$

Hence, by (2), we find for the transverse curvature

$$\frac{1}{\rho'} = \frac{\theta}{2a} = \frac{x}{2ap} \text{ cm.}^{-1}. \quad \dots\dots\dots\dots\dots(3)$$

We have here supposed the axes of the steel needles to coincide with the faces of the rod, but it is easily seen that the result is the same when the axes of the needles are at small distances from the edges of the rod, as in Fig. 41. Further, there is no need for the needles to be absolutely straight.

The curvature $1/\rho$ of the neutral filament is deduced from the vertical motion of Q, the point midway between U and V. A pin is attached by wax to the rod at Q and a vertical scale (not shown in Fig. 40) is placed so that the tip of the pin moves along it when the rod is bent (see Fig. 35). If preferred, the mirror method described in § 67 may be used for finding ρ.

If the distance between the knife edges be $2l$ cm. and if h cm. be the distance through which Q rises when a mass M is placed in each pan, we have $l^2 = h(2\rho - h)$, or approximately, since h is small compared with ρ

$$\frac{1}{\rho} = \frac{2h}{l^2}. \quad \dots\dots\dots\dots\dots\dots\dots\dots(4)$$

75. Determination of Poisson's ratio. A series of observations is made. The masses in the pans are varied by equal steps from zero to some maximum value which does not strain the rod too much. In putting the masses into the pans care must be taken that one pan is not so heavily loaded that the heavier load overbalances the other. To avoid this disaster, equal masses may be put into the two pans simultaneously, using both hands. The masses must be put into the pans as *gently* as possible so as to avoid any chance of disturbing the clamping of the long steel needles.

For each value of the load, beginning with the pans empty, the reading of the needle VP on the scale T is taken as well as the reading on the vertical scale of the pin attached to the centre of the rod. If the pans be light, we may take the readings when the pans are empty as the zero readings.

To determine Poisson's ratio from the observed quantities we use the values of $1/\rho'$ and $1/\rho$ given by (3) and (4), and thus we find

$$\sigma = \frac{\rho}{\rho'} = \frac{l^2}{4ap} \cdot \frac{x}{h}. \quad \dots\dots\dots\dots\dots\dots(5)$$

The results of the observations may be shown graphically, h being taken as abscissa and x as ordinate; a straight line is then drawn by the aid of a thread (page 83) so as to pass as evenly as possible among the plotted points. The difference in the values of x, *as shown by this line*, for $h = 0$ and for some definite elevation h is taken as the best value of x for that value of h. These values of h and x are used in calculating σ by the formula (5).

76. Practical example. The observations may be entered as in the following record of an experiment made by G. F. C. Searle upon a steel rod about 0·3 cm. (⅛ inch) in thickness.

Width of bar at centre $= 2a = 2·48$ cm.   Hence $a = 1·24$ cm.

Thickness of bar $= 2b = 0·3$ cm.   Hence $b = 0·15$ cm.

Distance between knife edges $= 2l = 40$ cm.   Hence $l = 20$ cm.

Distance from knife edge to point of support of corresponding pan $= 30$ cm.

Distance from tip of pointer to centre of edge of rod $= p = 43·0$ cm.

To give a clear idea of the magnitudes of the two radii of curvature, the values of ρ and ρ' have been calculated by (4) and (3). Thus,

$$\rho = \frac{l^2}{2h} = \frac{20^2}{2h} = \frac{200}{h}, \qquad \rho' = \frac{2ap}{x} = \frac{2·48 \times 43}{x} = \frac{106·6}{x}$$

Load in each pan grammes	Reading of pin at centre cm.	Reading of pointer cm.	h cm.	x cm.	$\dfrac{x}{h}$	ρ cm.	ρ' cm.
0	11·00	8·77	—	—	—	∞	∞
500	10·75	8·81	0·25	0·04	0·160	800	2670
1000	10·51	8·84	0·49	0·07	0·143	408	1520
1500	10·29	8·88	0·71	0·11	0·155	282	970
2000	10·04	8·91	0·96	0·14	0·146	208	762
2500	9·80	8·95	1·20	0·18	0·150	167	592
3000	9·60	8·99	1·40	0·22	0·157	143	485

When the values of x and h were plotted, the best value of x/h was found to be 0·152, which is identical with the mean of the values of x/h given in the table.   Hence, by (5),

$$\text{Poisson's ratio} = \sigma = \frac{l^2}{4ap} \cdot \frac{x}{h} = \frac{20^2}{4 \times 1·24 \times 43·0} \times 0·152 = 0·285.$$

The value of $a^2/2b$ is $1·24^2/0·3$ or 5·13 cm.   The least value of ρ' is 485 cm. and this is nearly 100 times the value of $a^2/2b$.   Thus the conditions laid down in Chapter II, § 33 are fully satisfied.

EXPERIMENT 10.   **Determination of Young's modulus by non-uniform bending of a rod.**

77. Introduction.   In EXPERIMENT 6 the rod is loaded in such a way that the bending is uniform for the part of the rod between the knife edges.   We now consider the case of non-uniform bending presented by a rod supported at its ends and loaded at its centre. The general mathematical theory of this case is beyond the scope of the present book, but it so happens that the conditions under which it is easy to make experiments with a rod loaded at its centre are those which must be satisfied in order that the results of the partial mathematical treatment, which is given below, may be good approximations to the truth.

To obtain a working knowledge of these conditions, we shall consider a weightless rod of length l, fixed horizontally at one end B, and bent by a downward vertical force F applied to the other end C in the manner indicated in Fig. 43. Let P be a

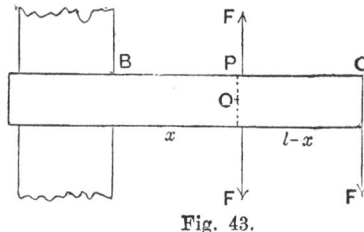

Fig. 43.

point on the rod at a distance x from the end B, let a transverse plane PO be drawn through P and let us consider the equilibrium of the portion PC. If we apply equal and opposite vertical forces, each equal to F, to the end of PC nearest to B, we see that the downward force F applied at C is equivalent to a downward force F applied at P together with a clockwise couple. Since the equilibrium of PC is maintained, the action of BP on PC is equivalent to an upward vertical force F together with a couple G in a direction tending to turn PC in a counter-clockwise direction, where

$$G = F(l - x). \qquad \dots\dots\dots\dots\dots\dots\dots(1)$$

The force F is supplied by the tangential stresses over the section, and hence, if V_{av} be the average vertical tangential stress,

$$A V_{\mathrm{av}} = F, \qquad \dots\dots\dots\dots\dots\dots(2)$$

where A is the area of the section.

The resultant of the normal stresses across the section vanishes, since the only force applied to PC is F, and F is vertical. Hence there is some point O in the section where the normal stress (T)

changes sign. Through O take rectangular axes OZ, perpendicular to the plane of the paper, and OY, perpendicular to the length of the rod, and take moments about OZ. Then, if $(Ty)_{av}$ be the average value of Ty, we have

$$A\,(Ty)_{av} = G = F(l - x). \quad \ldots\ldots\ldots\ldots\ldots\ldots(3)$$

From (2) and (3) we have

$$V_{av} = \frac{F}{A} = \frac{(Ty)_{av}}{l - x}.$$

We may conclude from this result that, when the greatest value of y is small compared with $l - x$, the average value of V is small compared with the values of T at the top and bottom of the rod, where the normal stress is greatest. Thus, we may expect the effects of the vertical shearing stress V, as shown in the deflexion of the end C, to be small compared with those of the normal stress T, provided the length of the rod be great compared with its depth*.

78. Approximate results for non-uniform bending.

When the vertical shearing stresses are neglected, we may neglect any changes in the angles of a square with horizontal and vertical sides in a plane parallel to that of the paper in Fig. 43, and hence a transverse section of the straight rod is strained into a surface cutting all the longitudinal filaments at right angles. If G be the point in the strained section which corresponds with the "centre of gravity" or centroid of the unstrained section, the plane touching the strained section at G will nowhere deviate appreciably from the strained section itself, provided the rod be thin, and consequently the assumption that the strained section is accurately plane will not lead to any appreciable error.

We suppose that the bending takes place parallel to a plane of symmetry of the rod, and, as in Chapter II, § 28, we call that plane the plane of bending. Corresponding to any transverse section, there is one longitudinal filament in the plane of bending which, in the neighbourhood of that section, remains unchanged

---

* Horizontal shearing stresses will act across the section through O (Fig. 43) as well as vertical shearing stresses. The effects of the former on the deflexion of C will be much less than that of the latter and are neglected in the investigation.

in length, and this filament is called the neutral filament corresponding to that section, while the straight line, which passes through the centre of curvature of the neutral filament at any point and is perpendicular to the plane of bending, is called the axis of bending for that point.

From the investigation of the uniform bending of a rod given in Chapter II, §§ 28 to 33, we may expect that, when the radius of curvature of the neutral filament is large enough compared with a quantity depending upon the form and magnitude of the transverse section, the tension of any longitudinal filament, at a point where the elongation is e cm. per cm., will not differ appreciably from that which would give the elongation e in a filament of equal unstrained section if the sides of this filament were free from stress. We shall therefore calculate the tension T in terms of Young's modulus E by the formula

$$T = Ee, \dots\dots\dots\dots\dots\dots\dots\dots\dots(4)$$

which, by Chapter I, § 17, applies to the case where the sides of the filament are free from stress.

If O (Fig. 44) be the point where the neutral filament cuts the plane of the transverse section, and if, in that plane, we take rectangular axes* OX, OY, parallel and perpendicular to the axis

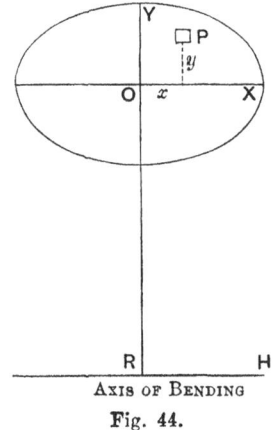

AXIS OF BENDING

Fig. 44.

* The axis of x is now taken at right angles to the plane of Fig. 43. In §§ 79, 81, 82 the axis of x is in the *same* direction as in Fig. 43.

of bending RH, it follows, as in Chapter II, § 29, that, if e be the elongation of the longitudinal filament through the point P, which has the coordinates x, y, then

$$e = y/\rho,$$

where ρ is the radius of curvature OR of the neutral filament. If T dyne cm.$^{-2}$ be the tension of the longitudinal filament, we have, by (4),

$$T = Ee = Ey/\rho.$$

We can now find the position of the neutral filament. For, if N be the total force acting across the transverse section and at right angles to it, and if α be an element of area, we have

$$N = \Sigma T\alpha = E\Sigma \alpha \dot{y}/\rho. \quad\dots\dots\dots\dots\dots\dots(5)$$

But $\Sigma \alpha y = Ah$, where h is the ordinate of the " centre of gravity " of the section, and A is the area of the section; thus

$$h = \Sigma \alpha y/A = \rho N/AE. \quad\dots\dots\dots\dots\dots(6)$$

Hence, when N is known, the position of the neutral filament relative to the " centre of gravity " of the transverse section is known.

In many cases the forces are applied to the rod in such a way that N is zero. In these cases h vanishes, and then the neutral filament passes through the " centre of gravity " of the section.

In the EXPERIMENT now under discussion the rod slides slightly over the knife-edges (Fig. 46) when the load is changed, and this motion is opposed by friction which therefore gives rise to a horizontal force. Since the depression of the centre of the rod due to a given load is found in practice to be nearly the same whether this load be reached by increasing a smaller load or decreasing a larger one (see §§ 68, 84), we may conclude that the effects of the horizontal force due to friction are small; in the present EXPERIMENT these effects will be neglected.

The sum of the moments, about the axis OX (Fig. 44), of the tensions in the longitudinal filaments is equal to the " bending moment," i.e. the moment about the same axis of the forces applied to the rod on either side of the transverse section. Hence, if the bending moment be G dyne-cm.,

$$G = \Sigma T\alpha y = \frac{E}{\rho}\Sigma \alpha y^2 = \frac{EI}{\rho}, \quad\dots\dots\dots\dots\dots(7)$$

where I is the "moment of inertia" of the section about the axis OX.

When N is zero or negligible, O may be taken to coincide with the centre of gravity of the strained section and when, in addition, the bending is slight, I may be taken as equal to I_0, the moment of inertia of the unstrained section about an axis through its centre of gravity parallel to the axis of bending.

From (6) and (7) we have

$$h = \frac{IN}{GA} = \frac{\rho N}{EA}, \quad \dots\dots\dots\dots\dots\dots(8)$$

and thus the distance of the neutral filament from the "centre of gravity" of the section will vary as we pass along the rod, unless N/G or ρN be constant. This condition is not generally satisfied, and hence, in general, there is no one longitudinal filament in the straight rod which suffers no elongation at every point of its length, when the rod is bent.

When a horizontal rod is bent by vertical forces, as in Fig. 46, the force N is negligible, but when a rod which is fixed at one end with its axis slightly inclined to the vertical is loaded at the other end, N will be large at all points of the rod, while G will be zero at the loaded end and will increase as the fixed end is approached. In this case h may be so large near the loaded end that the neutral filament does not lie within the rod; in other words, every longitudinal filament is either extended or shortened according to the direction of N. The moment of inertia I can then no longer be taken as equal to I_0 but must be found from the expression

$$I = I_0 + Ah^2, \quad \dots\dots\dots\dots\dots\dots\dots(9)$$

which is obtained in Note IV, § 12.

79. Cartesian expression for curvature. When a rod is bent by the application of known forces, the bending moment G is known at every point of the rod. When the effects of the force N are negligible, I may be taken as equal to I_0 and then equation (7) gives the curvature $1/\rho$, and from this information we have to determine the form of the rod. When, as in practical cases, the bending is slight, we can obtain a simple differential equation from which the form of the rod can be found, if we express $1/\rho$ in terms of Cartesian coordinates.

Let x, y be the coordinates of any point P (Fig. 45) on the curve AB. Let the radius of curvature at P be ρ, and let the tangent at P make an angle ψ with the axis of x. Then

$$\tan \psi = dy/dx. \quad \dots\dots\dots\dots\dots\dots(10)$$

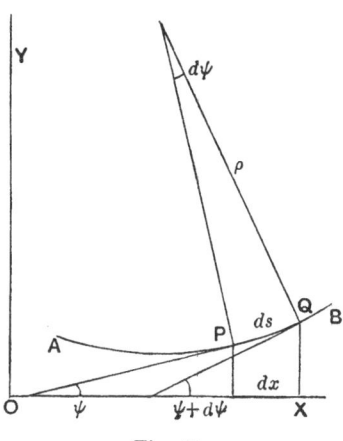

Fig. 45.

If the tangent at the neighbouring point Q make an angle $\psi + d\psi$ with OX, then $d\psi$ is the angle between the normals at P and Q, and hence, if the length of the element of arc PQ be ds, we find

$$dx/ds = \cos \psi, \quad \dots\dots\dots\dots\dots\dots(11)$$

and thus

$$\frac{1}{\rho} = \frac{d\psi}{ds} = \frac{d\psi}{dx} \cdot \frac{dx}{ds} = \cos \psi \frac{d\psi}{dx}. \quad \dots\dots\dots\dots(12)$$

By differentiating (10), we obtain

$$\frac{d^2y}{dx^2} = \sec^2 \psi \frac{d\psi}{dx},$$

and hence, by (12),

$$\frac{1}{\rho} = \cos^3 \psi \frac{d^2y}{dx^2} = \frac{d^2y}{dx^2} \Big/ \sec^3 \psi. \dots\dots\dots\dots(13)$$

But $\sec^2 \psi = 1 + \tan^2 \psi$, and hence

$$\frac{1}{\rho} = \frac{d^2y}{dx^2} \left\{ 1 + \left(\frac{dy}{dx} \right)^2 \right\}^{-\frac{3}{2}}. \quad \dots\dots\dots\dots(14)$$

When the curve is nearly parallel to OX, so that ψ is always very small, we may replace $\cos^3 \psi$ by unity and write

$$\frac{1}{\rho} = \frac{d^2 y}{dx^2} \quad \dots\dots\dots\dots\dots\dots\dots\dots(15)$$

80. Apparatus. The rod, of circular or rectangular section, rests upon two knife edges A, B (Fig. 46) carried by a stout bed XY such as is described in § 65. A light pan is suspended from C, the point of the rod midway between A and B. A pin is fixed to the rod by wax at C and a vertical scale is set up close

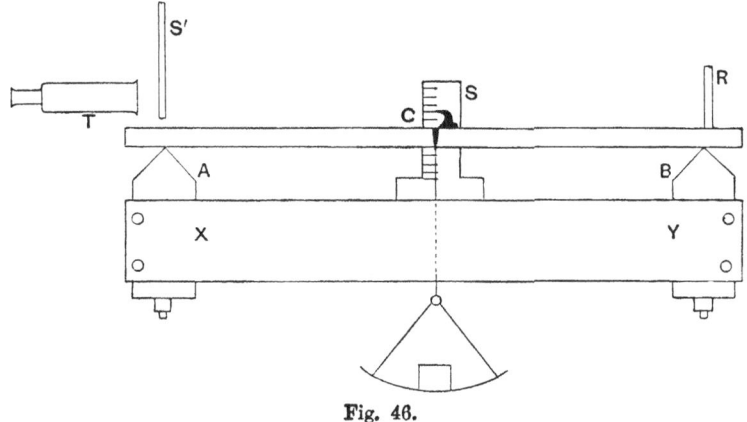

Fig. 46.

to the pin. To secure symmetrical loading, a stirrup, as in Fig. 47, may be used for suspending the pan from a rectangular rod. A plate perforated by a circular hole (Fig. 48) may be used with a round rod. A circular rod may be prevented from rolling on the knife edges by the device mentioned in § 65.

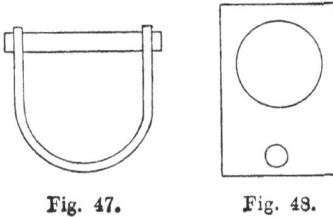

Fig. 47.   Fig. 48.

Instead of observing the depression of the central point C, we may observe the slope of the rod at B. A plane mirror R is attached to the rod immediately over the knife edge B and a vertical scale S' is set up over the knife edge A. By a telescope T fitted with cross-wires and held in a *firm* stand (§ 67), the scale S' can be viewed by reflexion at R; the slope of the rod at B can be deduced from the apparent motion of the scale past the horizontal cross-wire of the telescope.

81. Depression at centre of rod. Let P (Fig. 49) be any point on the rod and let the coordinates of P relative to the axes CX, CY through the central point C be x, y cm. Let

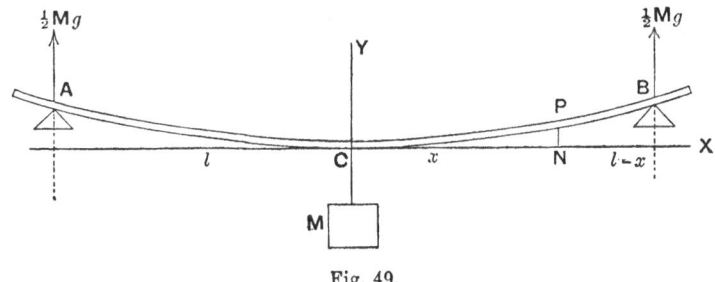

Fig. 49.

the distance between the knife edges A, B be $2l$ cm. For the sake of clearness, the curvature of the rod is greatly exaggerated in the figure. Let a load of M grammes be suspended at C. If the rod be so stiff that it is only slightly bent by its own weight, the additional depression due to M may be taken to be equal to the depression which M would cause if the rod were without weight.

In the case of a weightless rod, the upward force due to each knife edge is $\frac{1}{2}Mg$ dynes. Hence, if G be the bending moment at P, we have

$$G = \tfrac{1}{2}Mg\,(l - x) \text{ dyne-cm.} \quad \dots\dots\dots\dots(16)$$

Thus, by (7) and (16),

$$\tfrac{1}{2}Mg\,(l - x) = \frac{EI}{\rho}. \quad \dots\dots\dots\dots\dots(17)$$

or, by (15), $$\frac{d^2y}{dx^2} = \frac{1}{\rho} = \frac{Mg}{2EI}(l - x), \quad \dots\dots\dots\dots(18)$$

the differential equation from which the form of the rod is to be deduced.

Integrating (18) with respect to x, we have

$$\frac{dy}{dx} = \frac{Mg}{2EI}(lx - \tfrac{1}{2}x^2) + K, \quad \dots\dots\dots\dots(19)$$

where K is a constant. At C the rod is horizontal, and hence $dy/dx = 0$, when $x = 0$. Thus $K = 0$.

Integrating a second time, we find

$$y = \frac{Mg}{2EI}(\tfrac{1}{2}lx^2 - \tfrac{1}{6}x^3) + N,$$

where N is another constant whose value is zero, since the axes have been chosen so that $y = 0$ when $x = 0$. Hence the form of the rod is given by

$$y = \frac{Mg}{12EI}(3lx^2 - x^3). \quad \dots\dots\dots\dots\dots(20)$$

It should be noticed that this equation holds good only over the part CB of the rod. For points on AC, the bending moment is not $\tfrac{1}{2}Mg(l - x)$ but $\tfrac{1}{2}Mg(l + x)$, which leads to

$$y = \frac{Mg}{12EI}(3lx^2 + x^3). \quad \dots\dots\dots\dots\dots(21)$$

For values of x which are numerically equal but of opposite signs, equations (20) and (21) give identical values of y.

If the depression of the mid-point be h cm., we see that h is equal to the elevation of B above the axis CX. But at B, $x = l$ and hence, by (20),

$$h = \frac{Mgl^3}{6EI}, \quad \dots\dots\dots\dots\dots\dots(22)$$

and

$$E = \frac{Mgl^3}{6hI} \text{ dynes per square cm.} \quad \dots\dots\dots\dots(23)$$

When the rod is circular with diameter $2a$ cm. or radius a cm., we have, by Note IV, § 12,

$$I = \tfrac{1}{4}\pi a^4 \text{ cm.}^4,$$

and for a rectangular rod of width $2a$ and thickness $2b$ cm., the side whose length is $2b$ being vertical in the experiment,

$$I = \tfrac{4}{3}ab^3 \text{ cm.}^4$$

If ρ_0 be the radius of curvature at the mid-point, where $x = 0$, we have, by (18),

$$\rho_0 = \frac{2EI}{Mgl}. \qquad \dotfill (24)$$

Comparing (22) with (24), we find

$$\rho_0 = l^2/3h. \qquad \dotfill (25)$$

If the rod had been uniformly bent into an arc of radius ρ', the geometry of the circle would lead to the approximate result

$$\rho' = l^2/2h.$$

82. Slope at end of rod. Since $K = 0$ in (19), we find that, if ψ_B be the slope at B, where $x = l$,

$$\psi_B = \left(\frac{dy}{dx}\right)_{x=l} = \frac{Mgl^2}{4EI}. \qquad \dotfill (26)$$

If z be the distance through which the scale S' (Fig. 46) appears to move past the cross-wire of the telescope when the load M is placed in the pan, we have, as in § 67,

$$2\psi_B = z/2l.$$

Thus, by (26),

$$z = 4l\psi_B = \frac{Mgl^3}{EI}, \qquad \dotfill (27)$$

and

$$E = \frac{Mgl^3}{zI} \text{ dynes per square cm.} \qquad \dotfill (28)$$

We see from (22) and (27) that the distance through which the scale appears to move past the cross-wire is six times the distance through which the mid-point descends.

Equation (28) has been obtained on the usual assumption that the slope of the rod is everywhere so small that $\cos \psi$ may be replaced by unity in (13). But an exact expression for $\sin \psi$ is easily found. Since, by (12)

$$\cos \psi \frac{d\psi}{dx} = \frac{1}{\rho},$$

we have, by (17),

$$\cos \psi \frac{d\psi}{dx} = \frac{Mg}{2EI}(l - x).$$

On integrating this equation from $x = 0$ to $x = x$ and noting that $\sin \psi = 0$ when $x = 0$, we find

$$\sin \psi = \frac{Mg}{2EI}(lx - \tfrac{1}{2}x^2), \quad \ldots\ldots\ldots\ldots\ldots(29)$$

and thus

$$\psi_B = \sin^{-1}\left(\frac{Mgl^2}{4EI}\right) = \frac{Mgl^2}{4EI} + \frac{1}{6}\left(\frac{Mgl^2}{4EI}\right)^3 + \ldots\ldots\ldots\ldots(30)$$

When ψ_B is small, the value given by (26) does not differ appreciably from that given by (30).

83. Determination of Young's modulus. Young's modulus may be deduced either from the depression of the mid-point or from the apparent motion of the scale past the cross-wire of the telescope. In either case a series of observations is made. The mass in the pan is increased by equal steps from zero to some maximum value which does not strain the rod beyond the elastic limit, and the mass is then diminished to zero by the same steps.

If the elongation of the most highly strained filament is not to exceed $\frac{1}{1000}$ cm. per cm., it follows, by Chapter II, § 29, that ρ_0 must not be less than $1000a$, where $2a$ is the diameter of the rod if circular, or its thickness (measured vertically) if rectangular. Hence, by (25), h should not exceed $l^2/(3000a)$.

At each stage of the loading and unloading a reading of the pin (or of the cross-wire) is taken. The difference between the mean of the two readings for a given mass M and the mean reading when the pan is empty is taken as the value of h (or of z) which corresponds to M. As in § 66, we may leave the mass of the pan out of account.

If the values of h/M or of z/M prove to be nearly constant for different loads, the mean value of h/M or of z/M may be used for finding Young's modulus. When the irregularities are serious, the graphical method should be used. In this case, the values of M and of h or of z are plotted and a straight line is drawn as evenly as possible among the plotted points. The difference between the values of h (or of z), *as shown by this line*, for $M = 0$ and for some definite mass M, is taken as the best value of h (or z) for that mass. These values of M and of h or z are used in (23) or in (28).

84. Practical example. The observations may be entered as in the following record of an experiment made by Messrs G. F. C. Searle and D. L. H. Baynes upon the circular rod of steel which was used in § 68, where the bending was uniform.

Readings of screw-gauge for pairs of diameters at right angles,

| ·9614 | ·9580 | ·9592 | ·9610 | ·9612 | ·9637 |
| ·9643 | ·9623 | ·9565 | 9595 | ·9623 | ·9625 |

Mean reading 0·9610 cm. Correction for zero error 0·0006 cm.; to be added.

Mean diameter $=2a=0·9616$ cm. Radius $=a=0·4808$ cm.

Hence, moment of inertia of section $=I=\frac{1}{4}\pi a^4=0·04197$ cm.^{4}.

Distance between knife edges $=2l=90$ cm.

Observations were taken both for the depression (h) at the mid-point and for the apparent displacement (z) of the scale past the cross-wire of the telescope, using the mirror method.

Load M grammes	DIRECT METHOD			MIRROR METHOD		
	Scale reading cm.	Mean depression h cm.	$\dfrac{1000h}{M}$	Scale reading cm.	Mean Displacement z cm.	$\dfrac{1000z}{M}$
0	5·01	—	—	23·40	—	—
2000	5·39	0·375	0·1875	25·57	2·180	1·090
4000	5·74	0·730	0·1825	27·76	4·365	1·091
6000	6·10	1·085	0·1808	29·92	6·535	1·088
8000	6·45	1·435	0·1794	32·09	8·685	1·086
6000	6·10	—	—	29·96	—	—
4000	5·75	—	—	27·78	—	—
2000	5·39	—	—	25·60	—	—
0	5·02	—	—	23·41	—	—

Mean value of $h/M=1·826 \times 10^{-4}$ cm. grm.$^{-1}$.

Mean value of $z/M=10·89\times10^{-4}$ cm. grm.$^{-1}$.

Dividing the mean value of z/M by six, we have $1·815 \times 10^{-4}$, which agrees fairly with the mean value of h/M.

Using the direct method, we have, by (23),

$$E=\frac{Mgl^3}{6hI} = \frac{981 \times 45^3}{6\times1·826\times10^{-4}\times0·04197}=1·94\times10^{12} \text{ dynes per square cm.}$$

Using the mirror method, we have, by (28)

$$E=\frac{Mgl^3}{zI} = \frac{981 \times 45^3}{1·089\times10^{-3}\times0·04197}=1·96 \times 10^{12} \text{ dynes per square cm.}$$

These values of E are about 5 per cent. lower than that obtained in § 68.

EXPERIMENT 11. **Determination of rigidity by the torsion of a blade.**

85. Introduction. The uniform torsion of a blade, i.e. a strip of metal whose thickness is very small compared with its width, is considered in §§ 42 to 45, Chapter II. In § 45 it is shown that, if G_1 be the torsional couple required to twist one end of a blade of length l cm. through ϕ radians relative to the other end, and if n be the rigidity,

$$G_1 = \frac{16\,nab^3\phi}{3l} \text{ dyne-cm.,} \dots\dots\dots\dots(1) \cdot$$

where $2a$ cm. is the width and $2b$ cm. the thickness of the blade. From this equation n can be calculated, when the relation of G_1 to ϕ has been determined. In the following experiment the determination is made by a dynamical method.

Polished strips of tempered steel form suitable specimens for this experiment and for EXPERIMENT 12. Since they are fairly uniform in thickness, reasonably good measurements can be made upon them. They have the further great advantage that they may be subjected to considerable strains without sustaining any permanent set. Steel strips can be obtained from the manufacturers in a great variety of widths and thicknesses, down to a thickness of $\frac{1}{500}$ inch (0·00508 cm.). When not in use they should be coated with vaseline to prevent rusting.

86. Determination of rigidity. The dynamical method of EXPERIMENT 5 may be employed with a slight alteration in the form of the inertia bar. The blade is clamped between two inertia bars of rectangular section by the aid of three screws in the manner indicated in Fig. 50. The central screw passes through a hole in the blade but the other screws are sufficiently far apart to allow the blade to pass between them. A similar method of clamping may be employed for the upper end of the bar; in this case the bars between which the upper end of the blade is clamped must be secured to a firm support. The edges of the blade must be vertical and the axes of the inertia bars and of the clamping bars must be horizontal.

For the reasons given in Note VII, the mass of each inertia bar should be determined *before* the screw holes are bored in it; for convenience the masses should be stamped or engraved on the bars.

The length of the blade, l cm., between the clamping bars at its upper end and the inertia bars at its lower end is measured by aid of a centimetre scale, and the average width, $2a$ cm., of the

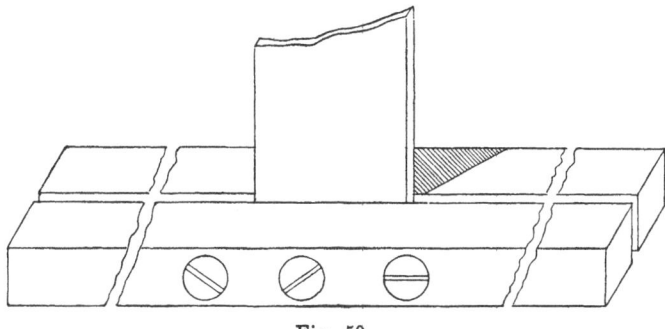

Fig. 50.

blade is found from a number of readings taken with a pair of sliding calipers. In measuring the thickness, $2b$ cm., at least 10 readings, at points fairly distributed over the blade, are taken with a screw-gauge and the correction for the zero error of the gauge is applied, with its proper sign, to the mean of the readings.

The lengths $2L_1$ and $2L_2$ cm. of the two inertia bars are then found; their widths $2A_1$ and $2A_2$ cm., measured at right angles to the plane of the blade, are also determined. Let the masses of the bars be M_1 and M_2 grammes. Then, if M_1, L_1 and A_1 be nearly equal to M_2, L_2 and A_2, we may treat the inertia system as if it were built up of two equal bars, each having the constants M, L and A, where

$$M = \tfrac{1}{2}(M_1 + M_2),\ L = \tfrac{1}{2}(L_1 + L_2),\ A = \tfrac{1}{2}(A_1 + A_2).$$

The moment of inertia of one of these bars about an axis through its centre of gravity parallel to the axis of the blade is, by Note IV, § 6, $\tfrac{1}{3}M(L^2 + A^2)$. The distance between these two axes is $A + b$ and hence, by the theorem of parallel axes (Note IV, § 3), the moment of inertia of each bar about the axis of the blade is

$$\tfrac{1}{3}M(L^2 + A^2) + M(A + b)^2.$$

Taking account of both bars, we see that, if the moment of inertia of the inertia system be K grm. cm.²

$$K = 2M(\tfrac{1}{3}L^2 + \tfrac{4}{3}A^2 + 2Ab + b^2).$$

In practice the last two terms in the bracket will be negligible in comparison with $\tfrac{1}{3}L^2$ and thus we may write

$$K = \tfrac{2}{3}M(L^2 + 4A^2)\,\text{grm. cm.}^2 \quad \ldots\ldots\ldots\ldots(2)$$

To complete the observations, the periodic time of the vibrations of the inertia system about the vertical axis of the blade must be determined; the arc of vibration should be small, since the theory of Chapter II, §§ 42 to 45, is only applicable to very small values of the twist per cm. At least two observations of the periodic time are made exactly as in EXPERIMENT 5, to which the reader is referred. Let the periodic time be T seconds.

By Note III, § 2, the angular acceleration of the inertia system is equal to G_1/K and hence, by (1), has the value

$$\frac{16nab^3\phi}{3lK} \text{ radians per sec. per sec.}$$

Hence, by Note V, § 2, the periodic time is given by

$$T = \frac{2\pi}{\sqrt{\text{angular acceleration for one radian}}}$$

$$= 2\pi\sqrt{\frac{3lK}{16nab^3}} \text{ seconds.}$$

Hence we obtain

$$n = \frac{3\pi^2 lK}{4ab^3 T^2} \text{ dynes per square cm.} \ldots\ldots\ldots\ldots(3)$$

From this equation the rigidity is determined.

87. Practical example. The observations may be entered as in the following record of an experiment on a blade of tempered steel.

Length of blade under torsion $= l = 58{\cdot}07$ cm.

The thickness was measured by a screw-gauge at 11 equidistant points along each of two lines parallel to the edges of the blade, the distance between each line and the nearer edge being about one-third the width of the blade. The following pairs of readings were obtained; they are expressed in hundredths of a centimetre.

4·66	4·70	4·68	4·67	4·66	4·70
4·80	4·78	4·79	4·75	4·74	4·78

4·70	4·66	4·71	4·68	4·66	mean $4{\cdot}680\times10^{-2}$ cm.
4·81	4·77	4·80	4·81	4·80	mean $4{\cdot}784\times10^{-2}$ cm.

The mean of the two means is 4.732×10^{-2} cm. ; the zero correction 0.08×10^{-2} cm. is to be added.

Hence thickness $= 2b = 4.812 \times 10^{-2}$ cm. Thus $b = 2.406 \times 10^{-2}$ cm.

Readings of sliding calipers on blade

5·06, 5·05, 5·06, 5·05, 5·02, 5·02

5·01, 5·02, 5·05, 5·06, 5·06. Mean 5·042 cm.

Zero error negligible.

Hence width $= 2a = 5.042$ cm. Thus $a = 2.521$ cm.

Masses of inertia bars: $M_1 = 796$, $M_2 = 796$. Hence $M = 796$ grammes.

Lengths of inertia bars: $2L_1 = 60.00$, $2L_2 = 60.00$. Hence $L = 30.00$ cm.

Widths of inertia bars: $2A_1 = 1.26$, $2A_2 = 1.26$. Hence $2A = 1.26$ cm.

Moment of inertia of system $= K = \tfrac{2}{3} M (L^2 + 4A^2)$

$$= \tfrac{2}{3} \times 796 \,(900 + 1.6) = 4.784 \times 10^5 \text{ grm. cm.}^2$$

Time of 50 complete vibrations: 128·6, 128·8, 128·7. Mean 128·7 secs.

Hence $T = 2.574$ secs.

Thus, by (3)

$$\text{Rigidity} = n = \frac{3\pi^2 l K}{4ab^3 T^2} = \frac{3\pi^2 \times 58.07 \times 4.784 \times 10^5}{4 \times 2.521 \times 0.02406^3 \times 2.574^2}$$

$$= 8.839 \times 10^{11} \text{ dynes per square centimetre.}$$

On account of the uncertainty as to the thickness of the blade, this result is uncertain to the extent of 4 or 5 per cent. The figures yielded by the logarithmic computation are, however, retained for use in connexion with EXPERIMENT 12.

EXPERIMENT 12. **Determination of $E/(1 - \sigma^2)$ by the uniform bending of a blade.**

88. Introduction. The uniform bending of a blade has been discussed in Chapter II and in § 37 of that chapter it is shown that, unless the bending be very slight, the bending moment, G_2, required to bend the blade so that the longitudinal filaments have a radius ρ cm., is given by

$$G_2 = \frac{4\,ab^3 E}{3(1 - \sigma^2)\rho} \text{ dyne-cm.,} \quad \dots\dots\dots\dots\dots\dots(1)$$

where $2a$ cm. is the width and $2b$ cm. is the thickness of the blade. Further, E is Young's modulus and σ is Poisson's ratio.

Equation (1) is not sufficient by itself to determine either E or σ, but if we use it in connexion with the equation (1) of § 85, viz.

$$G_1 = \frac{16\,nab^3 \phi}{3l} \text{ dyne-cm.,}$$

which refers to the torsion of the blade and with the equation (11) of § 19, Chapter I, viz.

$$E = 2n(1 + \sigma), \dots\dots\dots\dots\dots\dots(2)$$

which expresses the relation between Poisson's ratio and the two elastic constants E and n, we have sufficient equations to determine the three quantities E, n and σ.

89. Apparatus. The relation between G_2, the bending moment, and ρ, the radius of curvature of the longitudinal filaments, may be investigated by the method of EXPERIMENT 6, if we introduce slight modifications rendered necessary by the flexibility of the blade.

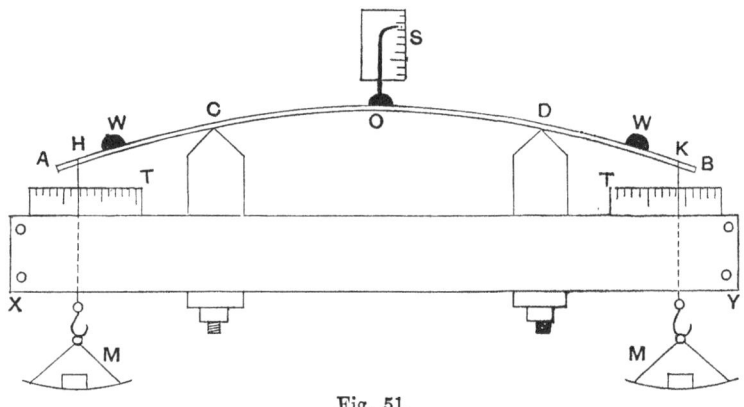

Fig. 51.

The blade AB (Fig. 51) rests symmetrically upon the knife-edges C, D, which are set at right angles to the length of the bed XY. The distance between the knife-edges should be adjusted so that the part of the blade between them is as nearly as possible straight when the blade is unloaded. When this adjustment is secured, it will be found that the distance between the knife-edges is approximately half the whole length of the blade*.

---

* It is easily seen that the bending moment at the *centre* of the unloaded blade is zero when the distance between the knife-edges is half the whole length of the blade. It can be shown that the centre of the blade is then raised above the level of the knife-edges by $h/80$, where h is the depression at the centre when the blade is supported at its *ends*. If the adjustment be such that the centre of the unloaded

The deflexions of the central point O are observed by aid of the scale S, the point of a bent needle serving as an index. A telescope may be used to magnify the scale and to avoid parallax.

The curvature may also be determined by the mirror method (EXPERIMENT 6, §67). In this case the curvatures employed may be smaller than when a needle-point and scale are used.

The pans for carrying the weights may be hung from threads which pass, at H and K, through the holes drilled in the blade to accommodate the central clamping screws (Fig. 50) used in EXPERIMENT 11. The threads may be secured to the blade by small pieces of wax W, W (Fig. 51) and should be provided with light hooks to carry the pans. It is convenient to adjust the mass of each pan to some definite value, say 10 grammes.

The length AB is limited by the condition that the blade should not bend much under its own weight and hence the curvature must be considerable if the deflexion of the central point O is to be large enough for accurate observation with simple apparatus. But, when the blade is bent, the horizontal distance between the threads is less than when the blade is straight and thus a correction becomes necessary. To determine this correction, two horizontal scales T, T are placed close to the threads and the scale-readings of the threads are taken for each load. The zero readings should, strictly, be taken when the threads carry no loads, but the threads would not be straight under those conditions, and thus the readings obtained when only the pans hang from the threads are treated as the zero readings; the horizontal displacements of the threads due to the pans alone are negligible.

On account of the bending, the blade must slide slightly on the knife-edges, and small differences in the friction between the blade and the knife-edges will be sufficient to cause the horizontal movement of one thread to differ from the horizontal movement of the other. The difference may be reduced, if necessary, by moving the blade through a small distance parallel to its length.

---

blade is at the same level as the knife-edges, the theory shows that the distance between the knife-edges is $6 - \sqrt{30}$ or ·5228 times the whole length of the blade. The radius of curvature at the centre of the blade is then $\frac{1}{2}R/(0\cdot5228 - \frac{1}{2})$ or $21\cdot9R$, where R is the radius of curvature at the centre when the blade is supported at its *ends*. It is impossible to secure entire freedom from bending.

90. Determination of $E/(1 - \sigma^2)$. It follows from §66, equation (1) (EXPERIMENT 6), that, when the horizontal distance of each thread from the nearer knife-edge is p cm., the bending moment, G_2, due to equal loads of M grammes hung from the threads is given by

$$G_2 = Mpg \text{ dyne-cm.} \dots\dots\dots\dots\dots\dots\dots(3)$$

On account of the unequal sliding of the blade on the knife-edges, the horizontal distances of the threads from the corresponding knife-edges are not quite equal. But, since the difference of distance is small, it will be sufficient to treat the system as symmetrical and to use for p in equation (3) the *average* distance of a thread from the corresponding knife-edge. If the distance, measured on the straight blade, between the points from which the threads are hung, be $2q$ cm., and if the displacements of the threads from their zero positions *towards* the knife-edges, as measured by the scales T, T, be t_1 and t_2 cm., the value of p is given by

$$p = q - l - \tfrac{1}{2}(t_1 + t_2), \dots\dots\dots\dots\dots\dots(4)$$

where $2l$ is the distance between the knife-edges. It is supposed, of course, that the system is symmetrical when the pans are unloaded.

The zero reading on the scale S is taken when only the threads and hooks hang from the blade. The pans are then hung on and a reading is again taken. The observations are continued for a series of loads in the pan, and readings are taken and *recorded* both for increasing and for decreasing loads. The difference between the mean of the two readings for a given load and the mean reading when only the threads and hooks hang from the blade is taken as the elevation of the central point due to that load. The readings of the threads on the horizontal scales are taken for each load, and the value of t_1 for a given load is calculated from the mean of the two readings of the corresponding thread for that load, and similarly for t_2.

The equation connecting ρ, the radius of curvature, with h, the elevation of the central point, is

$$h(2\rho - h) = l^2$$

or

$$\rho = \frac{l^2}{2h} + \frac{h}{2}, \dots\dots\dots\dots\dots\dots\dots(5)$$

where $2l$ is the distance between the knife-edges.

In EXPERIMENT 6, we could neglect $\frac{1}{2}h$ in comparison with $l^2/2h$, but in the present experiment the student should examine the relative magnitudes of the quantities so as to be able to decide whether the term $\frac{1}{2}h$ may be neglected or whether it must be retained.

The thickness and the width of the blade are measured exactly as in EXPERIMENT 11. If that experiment has been already performed upon the *same* blade, it will not be necessary to re-measure the blade.

From (1) and (3) we have

$$\frac{E}{1 - \sigma^2} = \frac{3M\rho pg}{4ab^3} \text{ dynes per square cm.,} \dots\dots\dots\dots(6)$$

and thus the values of $M\rho p$ found by experiment may be expected to be nearly constant. The mean of the values obtained for the series of loads is calculated and from it the value of $E/(1 - \sigma^2)$ is found by equation (6). If there be serious irregularities, the graphical method described in § 66 (EXPERIMENT 6) should be employed.

91. Calculation of E and of σ. Let the value of $E/(1 - \sigma^2)$ obtained from (6) be denoted by J, so that

$$\frac{E}{1 - \sigma^2} = J.$$

Substituting for E from (2), we have

$$\frac{2n}{1 - \sigma} = J$$

or
$$\sigma = 1 - \frac{2n}{J}. \dots\dots\dots\dots\dots\dots(7)$$

Using this value of σ in (2), we obtain

$$E = 4n\left(1 - \frac{n}{J}\right). \dots\dots\dots\dots\dots(8)$$

On comparing equation (3) of § 86 with equation (6) of § 90, it will be seen that n/J and, therefore also, σ are independent of the values adopted for the width and thickness of the blade.

92. Practical example. The observations may be entered as in the following record of an experiment made upon the blade of tempered steel used in the experiment of § 87.

Mean width of blade $= 2a = 5 \cdot 042$ cm.   Thus $a = 2 \cdot 521$ cm.

Mean thickness of blade $= 2b = 4 \cdot 812 \times 10^{-2}$ cm.   Thus $b = 2 \cdot 406 \times 10^{-2}$ cm. These values are those used in § 87.

Distance between knife-edges $= 2l = 33 \cdot 0$ cm.   Thus $l = 16 \cdot 5$ cm.

Distance between the points of support of the threads when the blade is straight $= 2q = 59 \cdot 40$ cm.   Hence $q = 29 \cdot 7$ cm.

Hence, by (4), $p = 29 \cdot 7 - 16 \cdot 5 - \frac{1}{2}(t_1 + t_2) = 13 \cdot 2 - \frac{1}{2}(t_1 + t_2)$.

The readings for the elevation, h, for increasing and decreasing loads were taken on a scale divided to half-millimetres. The masses of the pans (10 grammes each) are included in the loads. The student must record the two readings for each load and also the readings of the threads on the horizontal scales.

Load grms.	Elevation h cms.	$\dfrac{l^2}{2h}$ cms.	$\frac{1}{2}h$ cms.	ρ cms.	$\dfrac{t_1 + t_2}{2}$ cms.	p cms.	$\dfrac{M\rho p}{100}$ grm. cm.^{2}
10	0·175	778·0	0·1	778·1	0·015	13·18	1026
20	0·360	378·2	0·2	378·4	0·045	13·16	996
30	0·540	251·2	0·3	251·5	0·070	13·13	991
40	0·715	190·4	0·4	190·8	0·155	13·04	995
50	0·890	153·0	0·4	153·4	0·190	13·01	998
60	1·060	128·4	0·5	128·9	0·255	12·94	1001
70	1·215	112·0	0·6	112·6	0·340	12·86	1013
80	1·370	99·4	0·7	100·1	0·425	12·77	1023

Mean value of $\frac{1}{100}M\rho p = 1005$ grm. cm.^{2} Thus $M\rho p = 1 \cdot 005 \times 10^5$ grm. cm.^{2} Hence, by (6),

$$\frac{E}{1-\sigma^2} = \frac{3\,(M\rho p)\,g}{4ab^3} = \frac{3 \times 1 \cdot 005 \times 10^5 \times 981}{4 \times 2 \cdot 521 \times 0 \cdot 02406^3}$$
$$= 2 \cdot 106 \times 10^{12} \text{ dynes per square cm.}$$

By the experiment of § 87 upon the same blade

$$n = 8 \cdot 839 \times 10^{11} \text{ dynes per square cm.}$$

But $J = E/(1-\sigma^2) = 2 \cdot 106 \times 10^{12}$ dyne cm.$^{-2}$, and thus

$$\frac{n}{J} = \frac{8 \cdot 839 \times 10^{11}}{2 \cdot 106 \times 10^{12}} = 0 \cdot 4197.$$

Hence, by (7), we find for Poisson's ratio,

$$\sigma = 1 - \frac{2n}{J} = 1 - 0 \cdot 8394 = \mathbf{0 \cdot 1606}.$$

Finally, by (8), we find for Young's modulus

$$E = 4n \left(1 - \frac{n}{J}\right) = 4 \times 8 \cdot 839 \times 10^{11} (1 - 0 \cdot 4197)$$

$$= 2 \cdot 052 \times 10^{12} \text{ dynes per square cm.}$$

The value of E depends on the thickness of the blade and is uncertain to the extent of 4 or 5 per cent. (see § 87).

EXPERIMENT 13. **Test of Lord Rayleigh's reciprocal relations.**

93. Introduction. Let forces be applied to any number of points of a system formed of one or more rigid or elastic or fluid bodies. The change of form of the system will, as a rule, be accompanied by changes of temperature, according to the principles of thermodynamics, and hence the work done by the forces will be represented partly by potential energy due to the change of height of the centre of gravity of the system, partly by the energy of elastic strain and partly by the thermal energy corresponding to the changes of temperature. Since the elastic constants depend to some extent upon the temperature, the final change of form will depend upon the manner in which the thermal energy is dealt with. But there will be a definite relation between the forces and the change of form in two cases, which we shall call adiabatic and isothermal.

In the adiabatic case, the heat which appears in each part of the system is supposed to remain there and not to escape to other parts of the system or to the surrounding bodies by conduction or radiation. It is obvious that it is impossible to secure these conditions in practice when we wish to study the change of form of the system due to the steady application of the forces.

In the isothermal case, the temperature is supposed to be maintained constant at every part of the system while the form of the system is changing. This condition can be approximately secured in practice by applying the forces so gradually that conduction and radiation prevent any appreciable changes of temperature.

In the adiabatic case, the changes of form will depend upon the adiabatic values of the elastic constants, but in the isothermal case upon the isothermal values. The results of Chapter I, § 22, show that, for given forces, the changes of form will be nearly the same in the two cases, if we are dealing with metals, but in the theoretical discussion the distinction between the two cases will be maintained.

We now pass on to consider the action of two sets of forces. We shall suppose that one set is first applied and that the other set is afterwards superposed.

In the adiabatic case, the changes of form which the second set produces, and also the accompanying changes of temperature, will be independent of the effects already produced by the first set of forces, provided these latter effects be small enough. Hence we may say that the final change of form is the resultant of those due to the two sets of forces acting separately, and that the change of temperature at any point of the system is the resultant of the changes which occur there when each set of forces acts separately.

In the isothermal case, the change of form, which the second set of forces produces, will be independent of the change already produced by the first set, provided that the latter change be small enough, and thus the final change of form is the resultant of those due to each set of forces acting separately. Further, the heat which must be given to any elementary volume of the system to keep the temperature constant, while the strain is changed, is proportional to the change of strain and is independent of any small strain already existing. Hence, the total amount of heat given to the system to keep the temperature constant, while both sets of forces are brought into operation, is the resultant of the amounts required when each set acts separately.

Similarly, if the forces of the first and second sets are very small, the effects due to a third set of small forces are independent of the first and second sets.

In the following investigation the system is supposed to be supported at a definite number of points by fixed supports, and we shall study the effects due to a force X applied at a point A of the system and a force Y applied at a point B. In addition to these forces, gravity will act upon the whole system. The forces X and Y and also the earth's attraction will call into play corresponding reactions at the points of support. When the changes of form due to gravity are small and the forces X and Y are small, the change of form due to X will be proportional to X

and will be the same as if neither gravity nor Y acted. Similarly, the change of form due to Y will be proportional to Y and the same as if neither gravity nor X acted.

94. Work done by forces. Let the forces X and Y, which act at A and B, have definite directions, and let x and y be the displacements of A and B *measured in the directions of X and Y* from the positions of those points when X and Y are both zero *and gravity continues to act*. Then, by the generalised form of Hooke's law, by which the effects of each force are proportional to the force and independent of the other force and of gravity,

$$x = aX + c_1 Y, \quad \ldots\ldots\ldots\ldots\ldots\ldots(1)$$
$$y = c_2 X + b Y. \quad \ldots\ldots\ldots\ldots\ldots\ldots(2)$$

Here a, b, c_1, c_2 are four constants which have one set of values in the adiabatic case and another set in the isothermal case.

We shall now consider two methods of applying the forces X and Y and shall calculate the work done by them in each method.

In the first method, the force at A is applied gradually, starting from zero and reaching its full value X, while the force at B is zero. At the end of this operation, the displacement of A is aX and that of B is $c_2 X$. The work done by the force at A during this stage is the product of the average force and the final displacement of A and thus is $\frac{1}{2}aX^2$; no work is done at B for the force there is zero. The force at B is now gradually applied, starting from zero and reaching its full value Y, while the constant force X continues to act at A. The displacement of A now increases from aX to $aX + c_1 Y$ and that of B from $c_2 X$ to $c_2 X + b Y$. The additional work done in this stage by the constant force X is $c_1 YX$, and the work done by the increasing force applied at B is the product of the average value of that force and the displacement bY, and is, therefore, $\frac{1}{2}b Y^2$. If the application of the forces X and Y causes the centre of gravity of the system to descend through a distance h_1, the work done by gravity is Mgh_1, where M is the mass of the system. Hence, if W_1 be the total work done by the forces and by gravity, we have

$$W_1 = \tfrac{1}{2}aX^2 + c_1 X Y + \tfrac{1}{2}b Y^2 + Mgh_1. \quad \ldots\ldots\ldots\ldots(3)$$

In the second method of applying the forces, the order of the application of the forces is simply reversed. If the work done be W_2, we find in a similar manner

$$W_2 = \tfrac{1}{2}aX^2 + c_2XY + \tfrac{1}{2}bY^2 + Mgh_2, \quad \ldots\ldots\ldots\ldots(4)$$

where h_2 is the distance through which the centre of gravity has descended.

These relations hold good in both the adiabatic and the isothermal cases, though the four constants a, b, c_1, c_2 have different values in the two cases.

When the adiabatic condition prevails, no heat enters or leaves any part of the system, and therefore W_1 and W_2 are the amounts of energy gained by the system when the forces X and Y are applied in the two methods. But, by § 93, the strain and the temperature at every point of the system are independent of the order of application of the forces, and thus the final state, and therefore the final energy, of the system, must be the same in the two cases. Since the strain is the same, the centre of gravity descends through the same distance, and thus $h_1 = h_2$, and then, since the energy is the same, it follows, from (3) and (4), that $c_1 = c_2$.

When the isothermal condition prevails and every part of the system is always at a constant temperature, heat must enter or leave the system in order to keep the temperature constant, and hence we cannot now say that W_1 and W_2 are the amounts of energy gained by the system when the forces X and Y are applied in the two methods. But, if Q_1 and Q_2 be the amounts of heat given to the system and E_1 and E_2 be the energy gained by the system, when the forces are applied in the two methods, we have

$$E_1 = W_1 + Q_1 = \tfrac{1}{2}aX^2 + c_1XY + \tfrac{1}{2}bY^2 + Mgh_1 + Q_1, \quad \ldots(5)$$

$$E_2 = W_2 + Q_2 = \tfrac{1}{2}aX^2 + c_2XY + \tfrac{1}{2}bY^2 + Mgh_2 + Q_2. \quad \ldots(6)$$

Now, by § 93, the final strain at every point of the system is independent of the order of application of the forces, and thus the order of application does not affect either the energy of elastic strain or the motion of the centre of gravity. Hence $h_1 = h_2$. Again, by § 93, the heat given to the system to keep the temperature constant is independent of the order of application of the forces, and thus $Q_1 = Q_2$. Since both the energy of elastic strain and the heat given to the system are independent of the order of application, so is also the total gain of energy. Hence $E_1 = E_2$. But $h_1 = h_2$ and $Q_1 = Q_2$ and thus, by (5) and (6) $c_1 = c_2$.

Since, in both the adiabatic and the isothermal cases, $c_1 = c_2$, we may write

$$c_1 = c_2 = c, \quad \dots\dots\dots\dots\dots\dots\dots(7)$$

and thus the work done by the forces (not including gravity) is

$$W = \tfrac{1}{2}aX^2 + cXY + \tfrac{1}{2}bY^2, \quad \dots\dots\dots\dots\dots(8)$$

and is independent of the order of the application of the forces.

We can now write the relations connecting the forces with the displacements in the forms

$$x = aX + cY, \dots\dots\dots\dots\dots\dots\dots(9)$$

$$y = cX + bY. \quad \dots\dots\dots\dots\dots\dots(10)$$

If we solve these equations for X and Y, we find

$$X = \frac{bx - cy}{ab - c^2}, \quad \dots\dots\dots\dots\dots\dots(11)$$

$$Y = \frac{-cx + ay}{ab - c^2}. \quad \dots\dots\dots\dots\dots\dots(12)$$

From the equations connecting X and Y with x and y Lord Rayleigh[*] has deduced three reciprocal relations.

95. Lord Rayleigh's reciprocal relations. The displacement of B produced by a force X applied at A is, by (10),

$$y_{Y=0} = cX, \quad \dots\dots\dots\dots\dots\dots\dots(13)$$

the suffix $_{Y=0}$ indicating that no force is applied at B.

Similarly, the displacement of A produced by a force Y applied at B is, by (9),

$$x_{X=0} = cY. \quad \dots\dots\dots\dots\dots\dots\dots(14)$$

Hence, when X in (13) is equal to Y in (14), we have $x_{X=0} = y_{Y=0}$. The result may be stated in words as follows:

---

[*] *Philosophical Magazine,* XLVIII, p. 452 (1874), or *Scientific Papers,* Vol. I, Art. 32.

First reciprocal relation. The displacement of B due to a force applied at A is equal to the displacement of A due to an equal force applied at B.

A second relation follows from (11) and (12). If a displacement x be given to A, the force required to hold B at rest, so that $y = 0$, is, by (12),

$$Y_{y=0} = - cx/(ab - c^2), \dots\dots\dots\dots\dots\dots(15)$$

and similarly, if a displacement y be given to B, the force required to hold A at rest, so that $x = 0$, is, by (11),

$$X_{x=0} = - cy/(ab - c^2). \quad \dots\dots\dots\dots\dots(16)$$

Thus, when x in (15) is equal to y in (16), we have $X_{x=0} = Y_{y=0}$. The result may be stated in words as follows:

Second reciprocal relation. If the point A be held fixed, while B receives a displacement, the force required at A is equal to that required to hold B fixed when A receives an equal displacement.

A third relation may be deduced from (9) and (10). Let $Y_{y=0}$ be the force which must be applied to B to keep it at rest, so that $y = 0$, or, in other words, let $- Y_{y=0}$ be the reaction at B when X is applied at A. Then, putting $y = 0$ in (10), we have

$$\frac{- Y_{y=0}}{X_{y=0}} = \frac{c}{b} . \quad \dots\dots\dots\dots\dots\dots(17)$$

Now let X be removed and let a force Y act at B. Then, putting $X = 0$ in (9) and (10), we see that the displacements of A and B are connected by the relation

$$\frac{x_{X=0}}{y_{X=0}} = \frac{c}{b}, \quad \dots\dots\dots\dots\dots\dots\dots(18)$$

and thus, by (17), $$\frac{- Y_{y=0}}{X_{y=0}} = \frac{x_{X=0}}{y_{X=0}}. \quad \dots\dots\dots\dots\dots\dots(19)$$

The result may be expressed in words as follows:

Third reciprocal relation. When a force is applied at A, and B is held fixed, the ratio which the reaction at B bears to the force at A is equal to the ratio which the displacement of A bears to the displacement of B, when a force acts at B while A is free from force.

It must be remembered that, as was specified in § 94, the displacements x and y are *measured in the directions of* X *and* Y *respectively.* They do not necessarily represent the *total* displacements of A and B.

96. Apparatus. The first and third reciprocal relations may be tested by experiments made on a long steel rod. It is important that the rod should be long, in order that fairly large displacements may be obtained without straining the rod beyond the elastic limit. The rod rests on two knife-edges C and D (Fig. 52) fastened to a stout bed. If the rod be round, it may be

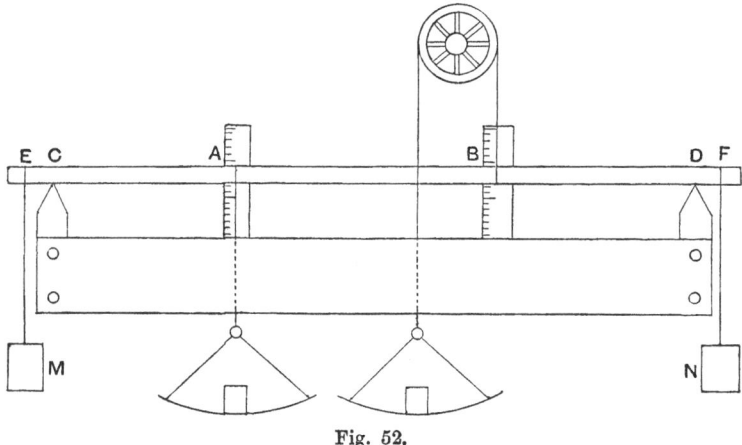

Fig. 52.

prevented from rolling by the device described in § 65. To make the arrangement available for testing the third reciprocal relation, masses M and N, each of two or three kilogrammes, are suspended from two points E and F on the rod close to the knife-edges. Pans are attached to the rod at A and B by hooks and strings. For testing the first relation both pans hang below the rod. For testing the third relation, the string carrying one pan passes over a pulley, care being taken that the part of the string between the hook and the pulley is vertical. To avoid errors due to friction, the pulley should be fitted with ball bearings, and the string should be flexible; *plaited* silk fishing line is suitable for the purpose. Pins are attached by wax to the hooks at A and B, and the displacements of these pins are found by means of two finely divided scales, which may conveniently be read by the aid of telescopes arranged to magnify them.

The apparatus is not well adapted for testing the second relation. When the B-pan is empty, a load of (say) a kilogramme at A will give that point a displacement large enough to be measured with some accuracy. But, if B be now brought back to its zero position by loads placed in the B-pan, the string passing over the pulley, it will be found that the displacement of A is greatly diminished. In the experiments described in § 99 the residual displacement was too small for accurate measurement.

Each pan produces a small displacement of A and B, but, if the pans remain in position during the experiment, these displacements are constant and the displacements due to the loads placed in the pans are simply added to these constant displacements. Hence we may neglect entirely the weights of the pans, provided the displacements due to the added loads be reckoned relatively to zero positions found when both the empty pans hang from the rod.

The positive directions of the forces X and Y will be taken to be vertically downwards; by § 94 the positive directions of x and y are also vertically downwards.

97. Test of first reciprocal relation. When the first relation is to be tested, both pans hang directly from the rod. The mass in the A-pan is increased from zero by equal steps, the B-pan being empty throughout, and the scale readings of A and B are taken at each stage. The B-pan is then loaded by equal steps, while the A-pan remains empty and the scale readings of A and B are again taken.

It will be found that the displacement of B due to any load at A is very nearly equal to the displacement of A due to an equal load at B, and thus the first reciprocal relation is verified.

From the first set of observations we obtain $x_{Y=0}$, and $y_{Y=0}$, the displacements of A and B due to the force X at A, and from the second set of observations we obtain $x_{X=0}$ and $y_{X=0}$, the displacements of A and B due to the force Y at B. From these quantities we can find a, b, c_1 and c_2. For, by (1) and (2),

$$a = \frac{x_{Y=0}}{X}, \qquad b = \frac{y_{X=0}}{Y}, \qquad c_1 = \frac{x_{X=0}}{Y}, \qquad c_2 = \frac{y_{Y=0}}{X}. \quad (20)$$

Each value of $x_{Y=0}$ in the first set of observations is divided by the corresponding force X and the mean value of $(x_{Y=0})/X$ is used for finding a. Similarly the mean value of $(y_{X=0})/Y$ gives b. The mean value of $(x_{X=0})/Y$ gives c_1 and the mean value of $(y_{Y=0})/X$ gives c_2. The agreement between c_1 and c_2 furnishes a test of the principles of energy employed in § 94. The mean of c_1 and c_2 may be taken as the value of c.

98. Test of third reciprocal relation. When the third relation is to be tested, the A-pan hangs directly from A but the string supporting the B-pan passes over the pulley. The mass in the A-pan is increased by equal steps and at each stage the load in the B-pan is adjusted until the scale reading of B is identical with the zero reading obtained when both pans are empty. Instead of attempting to make an exact adjustment of the load, we may take readings for two loads, one a little too great and the other a little too small, and may obtain the required load by interpolation.

The force due to the load in the A-pan is $X_{y=0}$; since the string supporting the B-pan pulls the rod upwards, the weight of the load in that pan is $-Y_{y=0}$. The value of $-Y_{y=0}/X_{y=0}$ is found for each load in the A-pan and the mean value is used for calculating c_2/b by the equation

$$c_2/b = -Y_{y=0}/X_{y=0}, \quad\quad\quad\quad\quad\quad (21)$$

which is derived from (2).

The experiments of § 97 give corresponding values of $x_{X=0}$ and $y_{X=0}$. From these the value of $x_{X=0}/y_{X=0}$ is found for each load in the B-pan and the mean value is used for calculating c_1/b by the equation

$$c_1/b = x_{X=0}/y_{X=0}, \quad\quad\quad\quad\quad\quad (22)$$

which is derived from (1) and (2).

The agreement between the values of c_2/b and c_1/b furnishes a test of the principles employed in § 94.

The pulley is now moved and the string supporting the A-pan is made to pass over it, while the B-pan is hung directly from B. A second set of observations is then made in which A is kept at rest. The mean value of $-X_{x=0}/Y_{x=0}$ derived from these observations is used to find c_1/a by the equation

$$c_1/a = -X_{x=0}/Y_{x=0}, \quad\quad\quad\quad\quad (23)$$

and this value is compared with the mean value of c_2/a derived from the experiments of § 97 by the equation

$$c_2/a = y_{Y=0}/x_{Y=0}. \quad \dots\dots\dots\dots\dots(24)$$

99. Practical example. The observations may be entered as in the following record of experiments made by G. F. C. Searle upon a steel rod 0·96 cm. in diameter and 160 cm. in length. The rod was supported on two knife-edges 140 cm. apart, and a mass of two kilogrammes was hung from each end of the rod as in Fig. 52. The point A was midway between the knife-edges, while B was 23·3 cm. from A. Scales divided to $\frac{1}{20}$ cm., on the sliders of two slide rules, were used in measuring the displacements and the readings were taken to $\frac{1}{200}$ cm. by aid of two telescopes. To avoid unnecessary complication, the forces were not measured in dynes but in terms of the *weight* of a kilogramme.

Test of first reciprocal relation. In the following tables, only the displacements, expressed in centimetres, are given, but the student must *record* all the readings and deduce the displacements from them.

Table 1.   *Table* 2.

X kilo. weight	$Y=0$				Y kilo. weight	$X=0$			
	$x_{Y=0}$ cm.	$y_{Y=0}$ cm.	$\dfrac{x_{Y=0}}{X}$	$\dfrac{y_{Y=0}}{X}$		$x_{X=0}$ cm.	$y_{X=0}$ cm.	$\dfrac{x_{X=0}}{Y}$	$\dfrac{y_{X=0}}{Y}$
0·5	0·330	0·275	0·660	0·550	0·5	0·270	0·255	0·540	0·510
1·0	0·665	0·565	0·665	0·565	1·0	0·565	0·520	0·565	0·520
1·5	1·005	0·845	0·670	0·563	1·5	0·845	0·775	0·563	0·517
2·0	1·340	1·130	0·670	0·565	2·0	1·120	1·030	0·560	0·515
2·5	1·665	1·410	0·666	0·564	2·5	1·415	1·305	0·566	0·522
3·0	1·995	1·695	0·665	0·565	3·0	1·695	1·560	0·565	0·520

Means   0·6660 0·5620   Means   0·5598 0·5173

The agreement between $y_{Y=0}$ and $x_{X=0}$ is satisfactory, the greatest difference being 0·01 cm. By (20), we find for the coefficients

$a=0·6660, \quad b=0·5173, \quad c_1=0·5598, \quad c_2=0·5620$ cm. per kilo. weight.

Hence c_1 and c_2 do not differ by as much as one part in two hundred and fifty. Taking the mean, we have

$$c=\tfrac{1}{2}(c_1+c_2)=0·5609.$$

The constancy of the ratios $x_{X=0}/y_{X=0}$ and $y_{Y=0}/x_{Y=0}$ can be examined in tables 3 and 4 which are derived from tables 2 and 1.

Table 3.

$X=0$

Y	0·5	1·0	1·5	2·0	2·5	3·0
$x_{X=0}/y_{X=0}$	1·059	1·087	1·090	1·087	1·084	1·087

Mean value 1·0823.

From the mean values given by table 2, we find

$$c_1/b = 0{\cdot}5598/0{\cdot}5173 = 1{\cdot}0822.$$

Table 4.

$Y=0.$

X	0·5	1·0	1·5	2·0	2·5	3·0
$y_{Y=0}/x_{Y=0}$	0·833	0·850	0·841	0·843	0·847	0·850

Mean value 0·8440.

From the mean values given by table 1, we find

$$c_2/a = 0{\cdot}5620/0{\cdot}6660 = 0{\cdot}8439.$$

In the ideal case the mean value of $x_{X=0}/y_{X=0}$ would be identical with

$$\frac{\text{mean value of } (x_{X=0}/Y)}{\text{mean value of } (y_{X=0}/Y)}$$

and similarly for $y_{Y=0}/x_{Y=0}$. The very small differences found in practice are due to the fact that, if P is the mean of the n quantities $p_1 \ldots p_n$ and Q is the mean of $q_1 \ldots q_n$, the mean of the quantities $p_1/q_1 \ldots p_n/q_n$ is not necessarily identical with P/Q unless $p_1/q_1 \ldots p_n/q_n$ are all equal.

Test of third reciprocal relation. The load in the A-pan was varied and the load in the B-pan was adjusted to make $y=0$, a variation of 10 grammes being just perceptible; then the load in the B-pan was varied and the load in the A-pan was adjusted to make $x=0$. The results are given in the following tables.

Table 5.   Table 6.

$X_{y=0}$ kilo. weight	$y=0$		$Y_{x=0}$ kilo. weight	$x=0$	
	$Y_{y=0}$ kilo. weight	$-\dfrac{Y_{y=0}}{X_{y=0}}$		$X_{x=0}$ kilo. weight	$-\dfrac{X_{x=0}}{Y_{x=0}}$
0·500	− 0·520	1·040	0·500	− 0·420	0·840
1·000	− 1·100	1·100	1·000	− 0·850	0·850
1·500	− 1·620	1·080	1·500	− 1·270	0·847
2·000	− 2·170	1·085	2·000	− 1·720	0·860
2·500	− 2·720	1·088	2·500	− 2·140	0·856

Mean 1·0786   Mean 0·8506

Thus, by (21) and (23),

$$\frac{c_2}{b} = \frac{-Y_{y=0}}{X_{y=0}} = 1\cdot0786, \quad \frac{c_1}{a} = \frac{-X_{x=0}}{Y_{x=0}} = 0\cdot8506.$$

From the mean values given in tables 3 and 4, we find, by (22) and (24),

$$\frac{c_1}{b} = \frac{x_{X=0}}{y_{X=0}} = 1\cdot0823, \quad \frac{c_2}{a} = \frac{y_{Y=0}}{x_{Y=0}} = 0\cdot8440.$$

The difference between c_2/b and c_1/b is about one part in 300 and the difference between c_1/a and c_2/a is about one part in 130.

EXPERIMENT 14. **Measurement of the energy dissipated through torsional hysteresis.**

100. Introduction. When a copper wire is subjected to gradually increasing torsion, the torsional couple is at first proportional to the twist, according to Hooke's law, but, as the twist is increased, the torsional couple fails to keep pace with the twist, and the ratio of the couple to the twist diminishes. If at any point in this later stage, we begin to reverse the twist, the torsional couple for a given angle is less during the untwisting than during the twisting, and hence less work is given out by the wire during the untwisting than was spent upon it during the twisting. If, starting with an untwisted wire, we first twist one end through an angle $+ \theta_0$ relative to the other, then reverse the motion till the twist is $- \theta_0$ and again reverse the motion till the twist is $+ \theta_0$, we shall find that the torsional couple when $+ \theta_0$ is reached for the second time, differs from the couple when $+\theta_0$ was reached for the first time. But if we subject the wire to many cycles of twisting and untwisting between the limits $+\theta_0$ and $-\theta_0$, we shall find that it settles down to a condition in which the couples called into play by the twists $+\theta_0$ and $-\theta_0$ have definite values, and that the couple called into play by any intermediate twist θ, has two definite values, one corresponding to the passage from $+\theta_0$ to $-\theta_0$ and the other to the passage from $-\theta_0$ to $+\theta_0$. When this condition is reached we say that the wire is in a cyclic state.

When the angle of torsion is large, there is a viscous yielding of the wire. Thus, if one end of the wire be suddenly twisted through a large angle relative to the other end, the torsional couple will not retain the value it has on the completion of the twist, but will diminish, at first rapidly and then more slowly, until, after some minutes, it has reached a steady value.

The viscous yielding of the wire makes it impossible to reach a steady state with cycles of torsion unless each cycle is performed in exactly the same manner, so that the time of passage from one angle to any other is the same for every cycle. Further, the work spent in taking the wire through a cycle with the given limits $+\theta_0$ and $-\theta_0$ will depend upon the speed at which cycles are performed.

When the cyclic state has been established, the twist will be related to the couple in the manner indicated in Fig. 53, where it

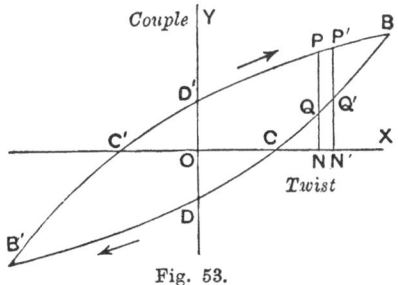

Fig. 53.

will be seen that the twist lags behind the couple. Thus, as we pass from B to B' the couple vanishes at C, but the twist does not vanish till we reach D. This lagging of the effect (the twist) behind the cause (the couple) has been called hysteresis by Ewing.

The phenomenon of hysteresis occurs in many other instances, one of the most important being the cyclical magnetisation of the magnetic metals. In the case of soft iron the resemblance is made specially close by the existence of magnetic viscosity.

101. Energy dissipated per cycle. We shall now show how to calculate the work spent in taking the wire through a cycle of torsion. At any time, when the twist is θ radians, let the couple be G dyne-cm. Then, when the twist increases by $d\theta$, the couple does $Gd\theta$ ergs of work*. If P, P' (Fig. 53) be two points on $B'D'B$ corresponding to θ and to $\theta + d\theta$, this work is represented by the number of units of area in the strip $PP'N'N$, provided that unit length along OX represents one radian and that unit length along OY represents a couple of one dyne-cm. If Q, Q' be the points on BDB' corresponding to the same angles, the angle *diminishes* by $d\theta$ as we pass from Q' to Q, and hence the wire gives out work represented by $QQ'N'N$. The resultant quantity of work spent upon the wire during the two changes is therefore represented by the area of the strip $PP'Q'Q$, and hence, if the work spent upon the wire during the complete cycle be W ergs, W is represented by the whole area $BDB'D'$.

In practice it would be inconvenient to plot the couple (measured in dyne-cm.) and the angle (measured in radians) upon the same scale. We shall therefore suppose that the scales are so chosen that one cm. (or one inch, if the squared paper be ruled in inches) along OX represents p radians, and that one cm. (or one inch) along OY represents q dyne-cm. Then the angle $d\theta$ is represented by $d\theta/p$ cm. (or inches) and hence the distance NN' is $d\theta/p$ cm. (or inches). Similarly, a couple G dyne-cm. is represented by G/q cm. (or inches) and thus the distance PN is G/q cm. (or inches). The area of the strip $PP'N'N$ is $Gd\theta/pq$ square cm. (or square inches) and hence the work $Gd\theta$ is pq times the area of the strip. Thus the work done during the cycle is now pq times the area of the whole curve or, in symbols,

$$W = pqA \text{ ergs}, \quad \dots\dots\dots\dots\dots\dots(1)$$

where A square cm. (or square inches) is the area of the curve as drawn on the paper.

* See Note VIII, equation (2).

102.   Apparatus.   A diagram of the apparatus is shown in
Fig. 54.   The copper wire A, about 0·1 cm. in diameter and 30 to
40 cm. in length, is soldered into a vertical rod B, which carries a
torsion head H moving past an index P.   The lower end of the
wire is soldered into one end of a short rod C.   Into the other end
of C is soldered a steel or brass wire D about 0·2 cm. in diameter

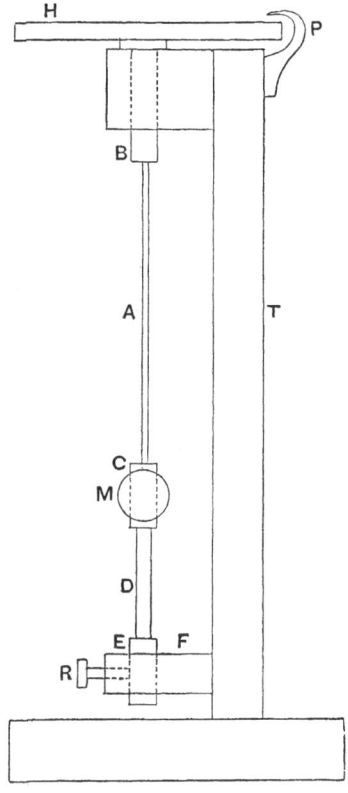

Fig. 54.

and about 10 cm. in length.   The lower end of D is soldered into
a rod E, which is secured in the block F by a clamping-screw R.
The block F and the bearing of the torsion head are fixed to an
upright T attached to a solid base.   The rod E is pulled downwards while the screw R is tightened so that the wire may be in a
state of tension.

The resistance to torsion of the wire D is so much greater than that of the copper wire that the couple exerted by the copper wire is unable to twist D beyond the elastic limit, and hence the angle turned through by the rod C is proportional to the torsional couple. Thus, if we measure the angle turned through by C and know the couple required to give D a twist of one radian, we can at once calculate the torsional couple.

The angle turned through by C is negligible compared with that turned through by the torsion head, and thus we may regard the latter angle as measuring the twist for the copper wire.

The angles turned through by C may be observed by aid of a mirror M attached to C and of a lamp and scale, or of a telescope and scale. If the distance of the scale from the mirror be d cm. and if the spot of light or the cross-wire of the telescope move over x cm. along the scale, when C turns through ϕ radians from the position corresponding to zero couple, then, for small angles,

$$\phi = x/2d \text{ radians.} \dots\dots\dots\dots\dots\dots(2)$$

In some cases it may be more convenient to measure the angles by means of a long pointer attached to C and moving over a horizontal scale. If the length of the pointer, measured from the axis of the wire, be h cm. and if the angle ϕ correspond to a displacement of x cm. along the scale, then, for small angles,

$$\phi = x/h \text{ radians.}$$

The couple required to give D a twist of one radian is easily found by a dynamical method. An auxiliary wire 40 to 50 cm. in length is cut from the same specimen as D and its ends are soldered into two short rods. One of these rods is held in a suitable firm clamp so that the wire is vertical and the other rod is secured to an inertia bar, exactly as in Fig. 33. The periodic time, T seconds, of the torsional vibrations of the bar is observed, and the moment of inertia, K gramme cm.^{2}, of the inertia bar is calculated from its mass and its dimensions exactly as in EXPERIMENT 5, § 63.

Let the free length of the auxiliary wire be l cm. and let that of the wire D be a cm., and suppose that the couple required to give D a twist of one radian is μ dyne-cm. Then the couple required to give the auxiliary wire a twist of one radian is $\mu a/l$ dyne-cm. and hence, by Note III, § 2, the angular acceleration of the inertia bar when its displacement is one radian is $\mu a/lK$ radian sec.$^{-2}$. Hence, by Note V, § 2,

$$T = 2\pi \text{ (angular acceleration for one radian)}^{-\frac{1}{2}} = 2\pi (\mu a/lK)^{-\frac{1}{2}}.$$

Thus
$$\mu = \frac{4\pi^2 lK}{aT^2}. \quad\dots\dots\dots\dots\dots\dots\dots(3)$$

Hence, if G dyne-cm. be the torsional couple corresponding to a deflexion of x cm., when a lamp and scale is used, we find, by (2) and (3),

$$G = \mu\phi = \frac{\mu x}{2d} = \frac{2\pi^2 lK}{adT^2} x \text{ dyne-cm.} \quad\dots\dots\dots(4)$$

103. Experimental details. Since simultaneous readings of the torsion head and of the spot of light are required, two observers are necessary. The torsion head is first turned into its zero position and the clamping screw R is slackened to free the wire from any torsional couple, and the screw is then tightened. The mirror, the lamp and scale and the focussing lens are then adjusted so that a sharp image of a cross-wire is formed on the scale near its centre. One observer, who may conveniently sit on a stool placed on the table, manipulates the torsion head while the other observes and records the scale reading of one edge of the image of the cross-wire. If a cycle with the limits $+200°$ and $-200°$ is to be studied, the torsion head is turned to $+200°$ and the scale reading of the cross-wire is taken. The head is then turned back to $+160°$ and the scale reading is again taken, and this process is continued by steps of $40°$ till $-200°$ is reached. The motion is then reversed and readings are taken at intervals of $40°$ till $+200°$ is reached. This constitutes the first cycle. But, to eliminate initial effects, a second and a third cycle are performed without any break in the process of observing. The cyclic state will be more quickly reached if the wire be put through a few cycles of twisting between $+200°$ and $-200°$ before the observations are taken.

To avoid confusion during the observations, the observers should prepare beforehand a blank table in which the readings may be entered. The table, when completed, should be similar to the table given in § 104.

When the wire is strained beyond the elastic limit, the couple it exerts is easily changed by vibration. The apparatus should therefore be kept as free as possible from vibration while the observations are in progress.

When the torsion of the wire is being increased in either direction, viscous effects will be observed, and hence the observations should be made at roughly equal intervals of time, and the readings should be taken immediately after the torsion head has been moved.

It will probably be found that the readings obtained in the third cycle are practically identical with those obtained in the second cycle. If this be the case, we may consider that the cyclic state has been established, and may use the readings in the third cycle to determine the work spent per cycle. Since the area of a hysteresis loop, such as is shown in Fig. 55, does not depend upon the position of the origin, it will be sufficient to plot the scale readings of the cross-wire against the readings of the torsion head, without working out the actual deflexion for each reading.

Smooth curves are drawn through the points corresponding to the two sides of the hysteresis loop and the area of the loop in square cm. (or in square inches, if the squared paper be ruled in inches) is then determined by the trapezoidal rule explained in Note IX.

Since the scale readings of the cross-wire are taken at equal intervals of angle, they may be employed directly in calculating the area by the trapezoidal rule. The distance on the squared paper corresponding to the difference of the two scale readings for each reading of the head is found, and these distances are added together and their sum is multiplied by the distance on the squared paper corresponding to the step in angle. The result is the area of the loop. A practical illustration is given in § 104.

If one cm. (or one inch) along the axis of angle correspond to m degrees, it corresponds also to p radians, where

$$p = m\pi/180, \ldots\ldots\ldots\ldots\ldots\ldots\ldots\ldots(5)$$

and if one cm. (or one inch) along the axis of couple correspond to a motion of the spot of light through n cm., it also corresponds to a couple q dyne-cm. where, by (4),

$$q = 2\pi^2 nlK/adT^2. \ldots\ldots\ldots\ldots\ldots\ldots(6)$$

Hence, by (1), the work spent per cycle is given by

$$W = pqA = \frac{mn\pi^3 l K A}{90 a d T^2} \text{ ergs.} \quad \dots\dots\dots\dots(7)$$

On account of the twisting of the stout wire by which the couples are measured, the angle of twist of the copper wire is not quite equal to the angle shown on the torsion head. The work spent upon the two wires is correctly given by the area of the loop if Hooke's law holds good for the stout wire. But since the stout wire is not strained beyond its elastic limit, the work spent upon it during a complete cycle is zero, and thus the area represents the work spent on the copper wire in each cycle.

The student who wishes to pursue the subject should obtain the hysteresis loops for a series of values for θ_0, such as 50°, 100°... and should then draw a curve showing how W depends upon θ_0.

104. Practical example. The observations may be entered as in the following record of an experiment made by Messrs G. F. C. Searle and W. Burton upon a copper wire about 0·09 cm. in diameter and 36·5 cm. in length. The wire, by which the couple was measured, was of brass and about 0·18 cm. in diameter.

Head Reading	Scale Reading First Cycle		Scale Reading Second Cycle		Scale Reading Third Cycle		Differences
degrees	centimetres		centimetres		centimetres		cms.
+200	27·4	27·3	27·3	27·3	27·3	27·2	0
+160	25·4	27·0	25·3	26·9	25·2	26·9	1·7
+120	23·5	26·6	23·5	26·5	23·4	26·4	3·0
+ 80	21·8	26·0	21·8	25·9	21·8	25·8	4·0
+ 40	20·4	25·2	20·4	25·2	20·4	25·2	4·8
0	19·4	24·4	19·3	24·3	19·3	24·3	5·0
− 40	18·4	23·2	18·4	23·2	18·4	23·2	4·8
− 80	17·7	21·8	17·7	21·8	17·7	21·8	4·1
−120	17·1	20·2	17·2	20·2	17·2	20·2	3·0
−160	16·6	18·4	16·7	18·4	16·8	18·4	1·6
−200	16·4	16·4	16·4	16·4	16·4	16·4	0
							sum = 32·0

Mass of inertia bar $=M=820$ grammes.

Length of inertia bar $=2L=37\cdot92$ cm.

Width of inertia bar $=2A=1\cdot6$ cm.

Moment of inertia of bar $=K=\frac{1}{3}M\,(L^2+A^2)=\frac{1}{3}\,820\,(18\cdot96^2+0\cdot8^2)$
$$=9\cdot84\times10^4\ \text{grm. cm.}^2.$$

Time of 50 complete vibrations 126·2, 126·0.   Mean 126·1 secs.

Periodic time $=T=126\cdot1/50=2\cdot522$ secs.

Length of auxiliary wire $=l=51$ cm.

Length of wire measuring couple $=a=9\cdot8$ cm.

Distance of scale from mirror $=d=65\cdot5$ cm.

The torsional couple G is, by (4), connected with the deflexion x by the equation

$$G=\frac{2\pi^2 lK}{adT^2}\,x=\frac{2\pi^2\times51\times9\cdot84\times10^4}{9\cdot8\times65\cdot5\times2\cdot522^2}\,x=2\cdot43\times10^4\times x\ \text{dyne-cm.} \quad\ldots(8)$$

After the torsion head had been turned through two or three cycles with the limits $+200°$ and $-200°$, the readings in the above table were taken.

The readings for the third cycle agreed so closely with those for the second cycle that the third cycle was taken as closely representing the cyclic state of the wire. In the third cycle there was a slight discrepancy between the two

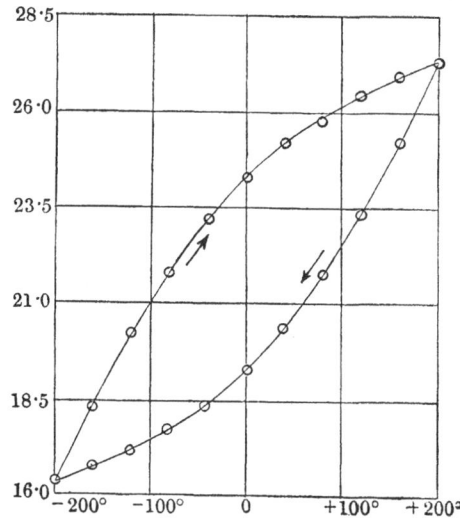

Readings of torsion head.

Fig. 55.

readings for $+200°$. The reading $27\cdot3$ was used in plotting the hysteresis loop. The differences corresponding to the two sides of the loop are given in the last column ; thus $26\cdot9 - 25\cdot2 = 1\cdot7$.

The loop shown in Fig. 55 was plotted on paper ruled in inches, and 2 cm. of deflexion and 80° of angle were each represented by one inch. The distances on the paper corresponding to the differences shown in the last column were therefore 0, $0\cdot85$, $1\cdot5$... inches, the sum being $16\cdot0$ inches. The step of 40° is represented by $\frac{1}{2}$ inch and hence A, the area, is $\frac{1}{2} \times 16\cdot0 = 8\cdot0$ square inches. Since one inch corresponds to 80°, $p = 80\pi/180 = 1\cdot396$ by (5), and since one inch corresponds to 2 cm. of deflexion, it also corresponds, by (8), to $2 \times 2\cdot43 \times 10^4$ dyne-cm. and thus q, the couple corresponding to one inch on the diagram, is $4\cdot86 \times 10^4$ dyne-cm. Hence, by (7), we find for the work spent per cycle

$$W = pqA = 1\cdot396 \times 4\cdot86 \times 10^4 \times 8\cdot0 = 5\cdot43 \times 10^5 \text{ ergs.}$$

NOTE I.

REDUCTION OF A GROUP OF FORCES TO A SINGLE FORCE AND A COUPLE.

Let any point O be taken as origin, let P be any other point, and let a force F act at P. Then apply to O (i) a force equal in magnitude to F and in the same direction, and (ii) an equal force in the opposite direction. The two forces of this pair are themselves in equilibrium and so have no resultant effect. The three forces can be replaced by a single force F acting at O in the same direction as the force at P together with a couple formed by the force at P and the remaining force at O. Treating all the other forces of the group in the same way, we see that the whole group is equivalent to a number of forces acting at O and to a number of couples. The forces may be combined into a single resultant force acting at O and the couples into a single resultant couple. The resultant force is clearly the same as if all the forces had acted at O in the first instance, but the magnitude of the couple will in general depend upon the position chosen for O.

NOTE II.

D'ALEMBERT'S PRINCIPLE.

Suppose that any particle of a solid or fluid body has an acceleration of f cm. sec. $^{-2}$ and that the mass of the particle is m grammes. Then the resultant of all the forces which act on the particle is the single force mf dynes in the direction of f. This force is called the "effective force."

The forces acting on the particle may be divided into two classes. The first class comprises the forces due to external bodies, whether they be transmitted by gravitational or electromagnetic action or are caused by the direct contact of some external body. The second class contains all those forces which act on the particle and are due to other particles of the body itself. These forces may arise from gravitational or electromagnetic action or from the direct contact with neighbouring particles. The resultant of these internal forces is a single force R dynes acting on the particle m.

But, by Newton's third law, the forces on any two particles due to their mutual action form a system in equilibrium*, and thus, when taken together, they have no component in any direction and no moment about any axis.

Hence, for any given body, the whole group of internal forces forms a system in equilibrium and gives rise to no force in any direction and no couple about any axis.

Now, if the force on the particle m due to external bodies be P dynes, the resultant of P and R is the "effective force" mf. Hence the system of applied forces and the system of internal forces are together exactly equivalent to the system of effective forces. But the internal forces form by themselves a system in equilibrium and therefore may be left out of account. We thus arrive at the result known as D'Alembert's Principle, which may be stated as follows :—

The system of "effective forces" is exactly equivalent to the system of applied forces, the resultants of the two systems having equal components in any direction and equal moments about any axis. (See Note I.)

Since the force mf dynes generates momentum in the direction of f at the rate of mf dyne-sec. per second, it follows that the rate at which the momentum of the whole body in any direction is increased is equal to the component in that direction of the system of applied forces. The C.G.S. unit of momentum is called a dyne-second because a dyne generates a unit of momentum in one second.

Again, since the rate of generation of momentum in the particle m is exactly represented by the force mf, the rate of increase of the moment of momentum or of the angular momentum of the particle about any fixed axis is exactly represented by the moment of the force mf about the same axis. Since the whole group of internal forces has no moment about any axis, it follows that the rate of increase of the angular momentum of the whole body about any fixed axis is equal to the moment about the same axis of the system of applied forces.

NOTE III.

MOTION OF A RIGID BODY.

1. ACCELERATION OF THE CENTRE OF GRAVITY. Let us take a set of rectangular axes fixed anywhere in space and let x_1, y_1, z_1 cm., x_2, y_2, z_2 cm. ... be the coordinates at time t of particles of masses m_1, m_2 ... grammes. Then, if ξ, η, ζ be the coordinates of the centre of gravity and M be the mass of the system of particles,

$$M\xi = \Sigma mx, \quad M\eta = \Sigma my, \quad M\zeta = \Sigma mz. \quad\text{......................}(1)$$

* This statement is no longer true when one or both of the particles is the source of electromagnetic radiation. In this case we have to consider forces acting on the ether itself.

If we denote by $\dot{x}_1$ the rate at which x_1 increases with the time, then $\dot{x}_1$ is the velocity of the particle m_1 in the positive direction of the axis of x. And if $\ddot{x}_1$ stand for the rate of increase of $\dot{x}_1$, then $\ddot{x}_1$ is the acceleration of m_1 in the same direction. Since, in the c.g.s. system, time is measured in seconds, the velocity is $\dot{x}_1$ cm. sec.$^{-1}$ and the acceleration is $\ddot{x}_1$ cm. sec.$^{-2}$. We shall extend this notation to the other coordinates. We then have, at once, for the velocities

$$M\dot{\xi} = \Sigma m\dot{x}, \quad M\dot{\eta} = \Sigma m\dot{y}, \quad M\dot{\zeta} = \Sigma m\dot{z}, \quad\dots\dots\dots\dots(2)$$

and for the accelerations

$$M\ddot{\xi} = \Sigma m\ddot{x}, \quad M\ddot{\eta} = \Sigma m\ddot{y}, \quad M\ddot{\zeta} = \Sigma m\ddot{z}. \quad\dots\dots\dots\dots(3)$$

Now, by Note II, $m_1\ddot{x}_1$ is the x-component of the effective force acting on the particle m_1. Since the whole group of internal forces has no component in any direction, it follows that $\Sigma m\ddot{x}$ is equal to the x-component of the whole group of applied forces. If the three components of the resultant of this group be X, Y, Z, we have $\Sigma m\ddot{x} = X$ and similarly for Y and Z. Hence, by (3),

$$M\ddot{\xi} = X, \quad M\ddot{\eta} = Y, \quad M\ddot{\zeta} = Z. \quad\dots\dots\dots\dots(4)$$

Thus $\ddot{\xi}$, $\ddot{\eta}$, $\ddot{\zeta}$ have exactly the same values as if the resultant of the applied forces acted on the whole mass collected into a single particle at the centre of gravity of the system. In other words :—

The acceleration of the centre of gravity of any system is the same as if the resultant of the applied forces acted on the whole mass collected into a single particle at the centre of gravity.

If F dynes be the resultant force, f cm. sec.$^{-2}$ the acceleration of the centre of gravity, and M grammes the mass of the system,

$$F = Mf.$$

This result is true for all systems of particles and is therefore true in the case of a rigid body.

2. Angular acceleration of a rigid body turning about a fixed axis. By Note II, the rate of increase of the angular momentum of any system about a fixed axis is equal to the moment about the same axis of the applied forces. When the system is a rigid body, the angular momentum (Note IV, § 13) is $K\omega$, where K grm. cm.^{2} is the moment of inertia of the body about the axis and ω radians per sec. is its angular velocity. If the rate of increase of ω be a radians per sec. per sec., then a is called the angular acceleration of the body. If the moment of the applied forces about the axis be G dyne-cm., it follows that

$$G = Ka,$$

since the quantity on the right side is the rate of increase of the angular momentum $K\omega$.

NOTE IV

MOMENTS OF INERTIA.

1. DEFINITION. As a knowledge of the moments of inertia of bodies of some simple forms is essential in practical work in elasticity, we give a sketch of the necessary propositions.

Let m_1, m_2 ... grammes be the masses of the particles of a rigid body and let r_1, r_2 ... centimetres be their perpendicular distances from a straight line or axis. Then the sum

$$m_1 r_1{}^2 + m_2 r_2{}^2 + \ldots = \Sigma m r^2$$

is called the moment of inertia of the body about the axis. We shall denote $\Sigma m r^2$ by K.

A system with unit moment of inertia is formed by a particle one gramme in mass placed at a distance of one centimetre from the axis, and the moment of inertia of this unit system is said to be one gramme-centimetre² or one grm. cm². If the moment of inertia of a body about any axis be K c.g.s. units or K grm. cm²., it has a moment of inertia K times as great as that of the unit system.

2. SOME PROPERTIES OF MOMENTS OF INERTIA. Take a set of rectangular axes OX, OY, OZ and let K_1, K_2, K_3 be the moments of inertia of the body about the three axes. If x, y, z be the coordinates of a particle of mass m, the square of its distance from OX is $y^2 + z^2$, and similarly for the other axes. Hence, by § 1, we have

$$K_1 = \Sigma m (y^2 + z^2), \quad K_2 = \Sigma m (z^2 + x^2), \quad K_3 = \Sigma m (x^2 + y^2). \ldots\ldots\ldots(1)$$

If the distance of m from O be R, and if H denote the sum $\Sigma m R^2$,

$$H = \Sigma m R^2 = \Sigma m (x^2 + y^2 + z^2).$$

By adding together the three equations (1) we find

$$K_1 + K_2 + K_3 = 2H. \ldots\ldots\ldots\ldots\ldots\ldots\ldots\ldots(2)$$

In some cases, such as that of a sphere with the origin at its centre, the three moments of inertia are equal; then

$$K_1 = K_2 = K_3 = \tfrac{2}{3} H. \ldots\ldots\ldots\ldots\ldots\ldots\ldots\ldots(3)$$

If the body be an infinitely thin plane lamina lying in the plane OXY, z is zero for every particle; then

$$K_1 = \Sigma m y^2, \quad K_2 = \Sigma m x^2, \quad K_3 = \Sigma m (x^2 + y^2) = K_1 + K_2. \ldots\ldots\ldots(4)$$

If the lamina be such that K_1 and K_2 are equal,

$$K_1 = K_2 = \tfrac{1}{2} K_3. \ldots\ldots\ldots\ldots\ldots\ldots\ldots\ldots\ldots(5)$$

3. THEOREM OF PARALLEL AXES. Let an axis passing through the centre of gravity of the body cut the plane of the paper at right angles at G (Fig. 56) and let any other parallel axis cut the paper at O. Let P be the projection of any particle (of mass m) on the plane of the paper and let PN be the perpendicular from P on OG. Let the mass of the body be M and let the moment of inertia of the body about the axis through G be K_0, and let that about the axis through O be K. Then

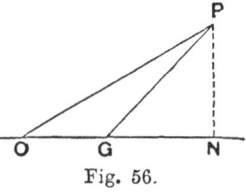

Fig. 56.

$$K_0 = \Sigma m \cdot PG^2,$$
$$K = \Sigma m \cdot OP^2 = \Sigma m \left(PG^2 + OG^2 + 2OG \cdot GN \right)$$
$$= \Sigma m \cdot PG^2 + OG^2 \Sigma m + 2 \, OG \, \Sigma m \cdot GN,$$

where GN is counted positive when N and O are on opposite sides of G. Since G is the projection of the centre of gravity, $\Sigma m \cdot GN = 0$ and hence

$$K = K_0 + M \cdot OG^2. \quad \dots\dots\dots\dots\dots\dots\dots\dots\dots\dots\dots(6)$$

Thus, when the moment of inertia, K_0, about any axis through the centre of gravity is known, the moment of inertia, K, about any parallel axis can be found at once by adding to K_0 the product of M and the square of the perpendicular distance between the two axes.

4. MOMENTS OF INERTIA OF A THIN UNIFORM ROD ABOUT ITS AXES OF SYMMETRY. Let the mass of the rod AB (Fig. 57) be M grammes and its length $2l$ centimetres. Let O be its middle point and let the axis of x coincide with OA. Since the rod is infinitely thin, $y = 0$ and $z = 0$ for every particle and thus, by (1), $K_1 = 0$. The moment of inertia about OY is proportional to l^2 when M

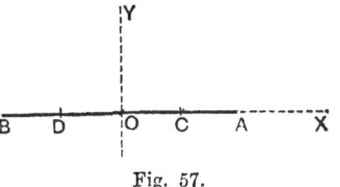

Fig. 57.

is given, for, if we uniformly stretch the rod to n times its original length, each particle will be n times as far from OY as it was originally, and therefore the new moment of inertia of each particle will be n^2 times its original value. Further, for a given length, K_2 is proportional to M. Thus we may put

$$K_2 = q \, Ml^2, \quad \dots\dots\dots\dots\dots\dots\dots\dots\dots\dots\dots(7)$$

where q is a numerical constant to be determined.

Now, by (7), the moment of inertia of the half rod OA about an axis through its centre of gravity C parallel to OY is $q \left(\tfrac{1}{2}M\right)\left(\tfrac{1}{2}l\right)^2$ or $\tfrac{1}{8}q \, Ml^2$. Hence, by § 3, the moment of inertia of the part OA about OY is

$$\tfrac{1}{8}q \, Ml^2 + \tfrac{1}{2}M \cdot OC^2,$$

and this is equal to $\tfrac{1}{2}K_2$, since the moment of inertia of OA about OY is half that of AB about the same axis. Hence, since $OC = \tfrac{1}{2}l$,

$$\tfrac{1}{2}q \, Ml^2 = \tfrac{1}{8}q \, Ml^2 + \tfrac{1}{8} \, Ml^2$$

or

$$q = \tfrac{1}{3}.$$

Thus, by (7),

$$K_2 = K_3 = \tfrac{1}{3} \, Ml^2. \quad \dots\dots\dots\dots\dots\dots\dots\dots\dots\dots(8)$$

5. MOMENTS OF INERTIA OF A UNIFORM RECTANGULAR LAMINA ABOUT ITS AXES OF SYMMETRY. Let the sides of the lamina be $2a$ and $2b$, and let its mass be M. Let O (Fig. 58) be its centre and let the axes OX, OY be parallel to the sides $2a$ and $2b$ respectively. Since K_1 is unchanged when the lamina is compressed into a uniform rod BB' lying along OY, we have, by (8),

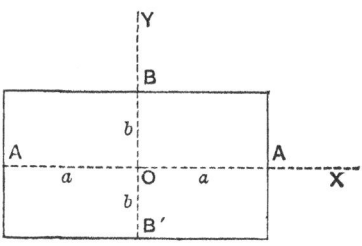

Fig. 58.

$$K_1 = \tfrac{1}{3}Mb^2. \quad\ldots\ldots\ldots\ldots\ldots\ldots\ldots\ldots\ldots\ldots\ldots\ldots(9)$$

Similarly

$$K_2 = \tfrac{1}{3}Ma^2. \quad\ldots\ldots\ldots\ldots\ldots\ldots\ldots\ldots\ldots\ldots\ldots(10)$$

By (4)

$$K_3 = K_1 + K_2 = \tfrac{1}{3}M(a^2 + b^2). \quad\ldots\ldots\ldots\ldots\ldots\ldots(11)$$

6. MOMENTS OF INERTIA OF A UNIFORM RECTANGULAR BLOCK ABOUT ITS AXES OF SYMMETRY. Take the origin O at the centre of the block and let OX, OY, OZ be parallel to the edges $2a$, $2b$, $2c$. Then K_1 is unchanged when the block is compressed into a uniform lamina in the plane OYZ and similarly for the other axes, and hence, by (11),

$$K_1 = \tfrac{1}{3}M(b^2 + c^2), \quad K_2 = \tfrac{1}{3}M(c^2 + a^2), \quad K_3 = \tfrac{1}{3}M(a^2 + b^2). \quad\ldots\ldots(12)$$

7. MOMENTS OF INERTIA OF A UNIFORM CIRCULAR LAMINA ABOUT ITS AXES OF SYMMETRY. Let the radius be a and the mass M, and let the axes OX, OY (Fig. 59) be in the plane of the lamina. Take a narrow strip PQP' parallel to OX, the points P, P' being on the circumference of the lamina and the point Q on OY, and let the mass of the strip be m. The moment of inertia of the strip about OX is $m . OQ^2$ and hence, by summation,

$$K_1 = \Sigma m . OQ^2.$$

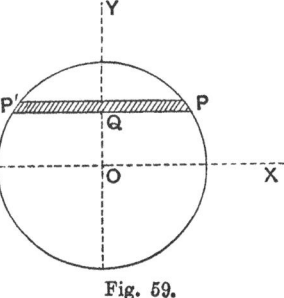

Fig. 59.

By (8), the moment of inertia of the *same* strip PQP' about OY is $\tfrac{1}{3}m . QP^2$ and hence, by addition,

$$3K_2 = \Sigma m . QP^2.$$

But, by symmetry, $K_1 = K_2$ and hence
$$4K_1 = K_1 + 3K_2 = \Sigma m \, (OQ^2 + QP^2) = \Sigma m \, a^2 = Ma^2.$$

Thus
$$K_1 = K_2 = \tfrac{1}{4} Ma^2. \quad\dots\dots\dots\dots\dots\dots\dots\dots\dots(13)$$

Then, by (4),
$$K_3 = K_1 + K_2 = \tfrac{1}{2} Ma^2. \quad\dots\dots\dots\dots\dots\dots\dots(14)$$

8. MOMENTS OF INERTIA OF A UNIFORM ELLIPTICAL LAMINA ABOUT ITS AXES OF SYMMETRY. Let the diameters of the ellipse, parallel to OX, OY be $2a$, $2b$. If, without change of mass, the circular lamina of § 7 be uniformly strained so that the point x, y is brought to the position ξ, η where $\xi = x$, $\eta = by/a$, the boundary will be an ellipse with diameters $2a$, $2b$. Further, if m be the mass of an element of the lamina,
$$K_1 = \Sigma m\eta^2 = (b^2/a^2) \Sigma my^2, \quad K_2 = \Sigma m\xi^2 = \Sigma mx^2.$$

But Σmx^2 and Σmy^2 are the same as K_2 and K_1 for the circular lamina, and hence, by (13), each is equal to $\tfrac{1}{4} Ma^2$. Thus, for the elliptical lamina
$$K_1 = \tfrac{1}{4} Mb^2, \quad K_2 = \tfrac{1}{4} Ma^2. \quad\dots\dots\dots\dots\dots(15)$$

By (4),
$$K_3 = K_1 + K_2 = \tfrac{1}{4} M (a^2 + b^2). \quad\dots\dots\dots\dots\dots(16)$$

9. MOMENT OF INERTIA OF A UNIFORM SOLID SPHERE ABOUT A DIAMETER. Let M be the mass and a the radius, and let the axes OX, OY, OZ pass through the centre O. Consider an element of the sphere in the form of a thin disk of radius r and of mass m with its plane parallel to the plane OXZ and at a distance y from it. Fig. 59 shows the section of the sphere and disk by the plane OXY. By § 7, the moment of inertia of the disk about that diameter of the disk which is parallel to OX is $\tfrac{1}{4}mr^2$ and hence, by the theorem of parallel axes (§ 3), its moment of inertia about OX is $\tfrac{1}{4}mr^2 + my^2$. Thus, by summation,
$$K_1 = \tfrac{1}{4} \Sigma mr^2 + \Sigma my^2.$$

Since $y^2 = a^2 - r^2$, we have
$$K_1 = \tfrac{1}{4}\Sigma mr^2 + \Sigma ma^2 - \Sigma mr^2 = Ma^2 - \tfrac{3}{4}\Sigma mr^2.$$

By § 7, the moment of inertia of the *same* disk about OY is $\tfrac{1}{2}mr^2$ and hence
$$K_2 = \tfrac{1}{2} \Sigma mr^2,$$
and thus
$$K_1 = Ma^2 - \tfrac{3}{2}K_2.$$

Since, by symmetry, $K_1 = K_2 = K_3$, we find that
$$K_1 = Ma^2 - \tfrac{3}{2}K_1.$$

Hence
$$K_1 = K_2 = K_3 = \tfrac{2}{5} Ma^2. \quad\dots\dots\dots\dots\dots\dots(17)$$

10. MOMENTS OF INERTIA OF A UNIFORM SOLID ELLIPSOID ABOUT ITS AXES OF SYMMETRY. Let the diameters parallel to OX, OY, OZ be $2a$, $2b$, $2c$. If, without change of mass, the sphere of § 9 be uniformly strained so that the point x, y, z is brought to ξ, η, ζ, where $\xi = x$, $\eta = by/a$, $\zeta = cz/a$, the sphere

will become the ellipsoid under consideration. If m be the mass of an element of volume,

$$\Sigma m\xi^2 = \Sigma mx^2, \quad \Sigma m\eta^2 = (b^2/a^2)\,\Sigma my^2, \quad \Sigma m\zeta^2 = (c^2/a^2)\,\Sigma mz^2.$$

But, by symmetry, $\Sigma mx^2 = \Sigma my^2 = \Sigma mz^2$ and, by § 2, each is equal to half the moment of inertia of the sphere about a diameter. Hence $\Sigma mx^2 = \frac{1}{5}\,Ma^2$. Thus

$$K_1 = \Sigma m\,(\eta^2 + \zeta^2) = \tfrac{1}{5}M\,(b^2 + c^2), \quad \dots\dots\dots\dots\dots(18)$$

$$K_2 = \Sigma m\,(\zeta^2 + \xi^2) = \tfrac{1}{5}M\,(c^2 + a^2) \quad \dots\dots\dots\dots\dots(19)$$

$$K_3 = \Sigma m\,(\xi^2 + \eta^2) = \tfrac{1}{5}M\,(a^2 + b^2). \quad \dots\dots\dots\dots\dots(20)$$

11. MOMENTS OF INERTIA OF A UNIFORM SOLID CIRCULAR CYLINDER ABOUT ITS AXES OF SYMMETRY. Let the mass of the cylinder be M, its length $2l$ and its radius a. Take the axis OX to coincide with the axis of the cylinder, and let O be the centre of the cylinder. Then the value of K_1 remains the same if the cylinder be compressed into a uniform circular lamina of radius a in the plane OYZ. Hence, by (14),

$$K_1 = \tfrac{1}{2}Ma^2. \quad \dots\dots\dots\dots\dots\dots\dots(21)$$

Now divide the cylinder into a series of infinitely thin disks by planes perpendicular to OX and let m be the mass of the disk which is at a distance x from the plane OYZ. The moment of inertia of this disk about a diameter is $\frac{1}{4}ma^2$, by (13), and hence, by the theorem of parallel axes (§ 3), its moment of inertia about OY is $\frac{1}{4}ma^2 + mx^2$. Thus, by summation,

$$K_2 = K_3 = \tfrac{1}{4}\Sigma ma^2 + \Sigma mx^2 = \tfrac{1}{4}Ma^2 + \Sigma mx^2.$$

But Σmx^2 is the moment of inertia about OY of a thin uniform rod lying along the axis of x and having the same mass and the same length as the cylinder; hence, by § 4, $\Sigma mx^2 = \frac{1}{3}Ml^2$. Thus

$$K_2 = K_3 = M\,(\tfrac{1}{4}a^2 + \tfrac{1}{3}l^2). \quad \dots\dots\dots\dots\dots(22)$$

12. "MOMENTS OF INERTIA" OF AREAS. If a be an element of any area and r be the perpendicular distance of a from a given axis, the quantity Σar^2 is called the "moment of inertia" or the second moment of the area about the axis; we shall denote it by I. The "moment of inertia" of the area is clearly equal to the moment of inertia of a uniform lamina of the same dimensions and of unit mass per unit area. The moment of inertia of the area may therefore be found by substituting A, the magnitude of the area, for M, the mass of the lamina. In the C.G.S. system, I will be expressed as a multiple of one cm⁴.

Theorem of parallel axes. If we apply the result of § 3 to an area, we see that if I_0 be the moment of inertia of an area A about an axis through its "centre of gravity," and I be the moment of inertia about a parallel axis,

$$I = I_0 + Ah^2,$$

where h is the perpendicular distance between the two axes.

From the results proved for laminas we obtain the following expressions:

Rectangular area of sides $2a$, $2b$. Here $A = 4ab$. Hence, by § 5, we find:—

About a diameter parallel to the side $2a$, $I_1 = \frac{1}{3}Ab^2 = \frac{4}{3}ab^3$

About a diameter parallel to the side $2b$, $I_2 = \frac{1}{3}Aa^2 = \frac{4}{3}a^3b$

About the normal through the centre, $I_3 = I_1 + I_2 = \frac{4}{3}ab\,(a^2 + b^2)$.

Circular area of radius a.   Here $A = \pi a^2$ and thus, by § 7, we have:—

About a diameter, $I_1 = I_2 = \frac{1}{4}Aa^2 = \frac{1}{4}\pi a^4$

About the normal through the centre, $I_3 = \frac{1}{2}Aa^2 = \frac{1}{2}\pi a^4$.

Elliptical area of diameters $2a$, $2b$.   Here $A = \pi ab$ and thus, by § 8, we have:—

About the diameter $2a$, $I_1 = \frac{1}{4}Ab^2 = \frac{1}{4}\pi ab^3$

About the diameter $2b$, $I_2 = \frac{1}{4}Aa^2 = \frac{1}{4}\pi a^3 b$

About the normal through the centre, $I_3 = I_1 + I_2 = \frac{1}{4}\pi ab\,(a^2 + b^2)$.

13. ANGULAR MOMENTUM OF A RIGID BODY TURNING ABOUT A FIXED AXIS. When a rigid body turns about a fixed axis, the velocity, and therefore also the momentum, of any particle is at right angles to the perpendicular, of length r cm., drawn from the particle to the axis. If the mass of the particle be m grammes and if the angular velocity of the body be ω radians per second, the velocity of the particle is $r\omega$ cm. sec.$^{-1}$ and its momentum is $mr\omega$ grm. cm. sec.$^{-1}$ or $mr\omega$ dyne-sec., a dyne-sec. being the amount of momentum which a dyne generates in one second. The moment of this momentum about the axis, i.e. the product of the momentum and the distance r, is $mr^2\omega$ grm. cm.^{2} sec.$^{-1}$ or $mr^2\omega$ dyne cm. sec. This is also called the angular momentum of the particle m about the axis. The angular momentum of the whole body is thus $\Sigma mr^2\omega$ or $\omega\Sigma mr^2$, since ω is the same for every particle because the body is rigid. The quantity Σmr^2 is K, the moment of inertia of the body about the axis. Hence:—

The angular momentum of a rigid body rotating with angular velocity ω *radians per sec. about a fixed axis is* $K\omega$ *grm. cm.^{2} sec^{-1}, where K grm. cm.^{2} is the moment of inertia of the body about the axis.*

14. KINETIC ENERGY OF A RIGID BODY TURNING ABOUT A FIXED AXIS. The kinetic energy of the particle in § 13 is $\frac{1}{2}m$ (velocity)^2 or $\frac{1}{2}mr^2\omega^2$ ergs, and thus, since ω is the same for every particle, the kinetic energy of the whole body is $\frac{1}{2}\omega^2\Sigma mr^2$ or $\frac{1}{2}K\omega^2$ ergs, where K grm. cm.^{2} is the moment of inertia about the axis.

NOTE V.

HARMONIC MOTION.

1. RECTILINEAR MOTION. On a circle with O (Fig. 60) for its centre take a point P and draw a perpendicular PM upon any diameter AOA'. Then, if P move round the circle with uniform speed, the point M moves along AOA'. The length OA is called the amplitude of the oscillation and the time occupied by M in going from A to A' and back to A is called the time of a complete vibration or the periodic time.

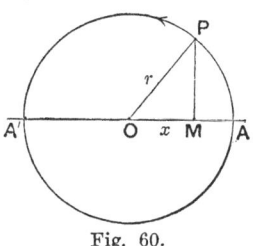

Fig. 60.

Let the radius of the circle be r cm. and let the speed of P along the arc of the circle be v cm. sec.$^{-1}$. If the angular velocity of OP be ω radians per second, the are described in one second is ωr cm. in length. Hence

$$v = \omega r. \quad \ldots\ldots\ldots\ldots\ldots\ldots\ldots\ldots\ldots\ldots\ldots\ldots(1)$$

Let the abscissa OM be x cm. and let the time t secs. be counted from the instant when P passes through A. Then the angle AOP is ωt radians, and hence

$$x = r \cos \omega t. \quad \ldots\ldots\ldots\ldots\ldots\ldots\ldots\ldots\ldots\ldots\ldots(2)$$

If the velocity of M along AOA' in the direction OA be u cm. sec.$^{-1}$, u is equal to the component, parallel to the same direction, of the velocity of P. Since the latter is at right angles to OP, we have

$$u = -v \sin POA = -\omega r \sin \omega t. \quad \ldots\ldots\ldots\ldots\ldots\ldots(3)$$

Since u is the rate at which x increases with the time, we see that the rate of increase of $r \cos \omega t$ is $-\omega r \sin \omega t$. Writing $\omega t + \frac{1}{2}\pi$ for ωt in these expressions and multiplying by ω, we see that the rate of increase of $\omega r \cos (\omega t + \frac{1}{2}\pi)$ is $-\omega^2 r \sin (\omega t + \frac{1}{2}\pi)$. But $\cos (\omega t + \frac{1}{2}\pi) = -\sin \omega t$ and $\sin (\omega t + \frac{1}{2}\pi) = \cos \omega t$, and thus the rate of increase of $-\omega r \sin \omega t$ is $-\omega^2 r \cos \omega t$ or $-\omega^2 x$. Hence, if the rate of increase of the velocity of M, i.e. the acceleration of M, be f cm. sec.$^{-2}$, we have

$$f = -\omega^2 x. \quad \ldots\ldots\ldots\ldots\ldots\ldots\ldots\ldots\ldots\ldots\ldots(4)$$

When x is positive, f is negative and *vice versa*, and thus f is always directed towards O.

As P goes round the circle, the point M oscillates along AOA', and the time of a complete vibration of M is equal to that of a complete revolution of P. Thus, if the periodic time be T seconds, the radius OP describes 2π radians in T seconds, and hence $\omega = 2\pi/T$, or

$$T = \frac{2\pi}{\omega}. \quad \ldots\ldots\ldots\ldots\ldots\ldots\ldots\ldots\ldots\ldots\ldots(5)$$

Since ω^2 is equal to the acceleration which M has towards O when M has unit displacement, i.e. when $x=1$, this result can be written

$$T = \frac{2\pi}{\sqrt{\text{acceleration for unit displacement}}}. \quad \dots\dots\dots(6)$$

If the acceleration of a given point moving along a straight line be proportional to its displacement from a fixed point on that line and be always directed towards that point, we can always find an auxiliary circle and an angular velocity such that the displacement, velocity and acceleration of M are equal to those of the given point, and hence the periodic time of the given point has the value stated in (6).

The motion of a point which vibrates so that its acceleration is proportional to its displacement from its mean position, is called *harmonic*. The radius of the auxiliary circle does not appear in the formula for the periodic time, and hence T is independent of the amplitude of the vibration. The vibrations are therefore called *isochronous*.

2. MOTION ABOUT A FIXED AXIS. In many cases of oscillation, the body, instead of moving along a straight line, turns about a fixed axis in such a way that its angular acceleration a is equal to $\mu\theta$ radians per sec. per sec., where θ radians is its angular displacement from its mean position and μ is a constant. If we now take a point M moving along AOA', as in Fig. 60, in such a way that OM is equal to $c\theta$, where c is a constant length, the acceleration of M will be equal to ca or to $c\mu\theta$, i.e. to $\mu \cdot c\theta$, and thus the acceleration of M is $\mu \cdot OM$. Hence, by § 1, the motion of M is harmonic, and T, the periodic time of its vibrations, is given by

$$T = \frac{2\pi}{\sqrt{\mu}}.$$

The angular motion of the body is said to be harmonic. Since $\sqrt{\mu}$ is equal to the angular acceleration of the body towards its mean position when its angular displacement is one radian, the last result can be written

$$T = \frac{2\pi}{\sqrt{\text{angular acceleration for one radian}}}.$$

NOTE VI.

CORRECTIONS FOR VARIATIONS IN THE RADIUS OF A WIRE.

1. YOUNG'S MODULUS. In finding Young's modulus by experiments on a wire of circular section it is usual to treat the wire as a circular cylinder with a radius equal to the mean radius of the wire, the latter being determined by observations at a number of points equally spaced along the wire. It may be useful to show how a closer approximation may be reached.

Suppose that the length L is divided up into m equal portions and that the radii measured at the centres of the first, second... portions are a_1, a_2.... Let a_0 be the mean radius and let

$$a_1 = a_0 + b_1, \quad a_2 = a_0 + b_2 \ldots .$$

Then, by the definition of mean radius,

$$\Sigma b = b_1 + b_2 + \ldots = 0.$$

To a close approximation we may treat the actual wire as if it were made up of m cylinders of length L/m and of radii a_1, a_2.... If l be the increase of length of the whole due to a longitudinal force F and if E be Young's modulus, we have, by equation (5), § 17, Chapter I,

$$l = \frac{FL/m}{\pi a_1^2 E} + \frac{FL/m}{\pi a_2^2 E} + \ldots$$

$$= \frac{FL}{m\pi E} \left\{ \frac{1}{(a_0 + b_1)^2} + \frac{1}{(a_0 + b_2)^2} + \ldots \right\}.$$

Expanding the m denominators by the binomial theorem, we have

$$l = \frac{FL}{m E} \left\{ \frac{1}{a_0^2} - \frac{2b_1}{a_0^3} + \frac{3b_1^2}{a_0^4} + \ldots + \frac{1}{a_0^2} - \ldots \right\}$$

$$= \frac{FL}{m\pi E} \left\{ \frac{m}{a_0^2} - \frac{2}{a_0^3} \Sigma b + \frac{3\Sigma b^2}{a_0^4} - \ldots \right\}.$$

The second term within the brackets vanishes since $\Sigma b = 0$. Hence, as far as the first correcting term,

$$l = \frac{FL}{\pi E a_0^2} \left\{ 1 + \frac{3\Sigma b^2}{m a_0^2} \right\}$$

and

$$E = \frac{FL}{\pi a_0^2 l} \left\{ 1 + \frac{3\Sigma b^2}{m a_0^2} \right\}. \quad \ldots\ldots\ldots\ldots\ldots\ldots\ldots\ldots(1)$$

Thus the value of E obtained by treating the wire as a cylinder of radius a_0 is slightly too small.

2. RIGIDITY. If we apply to equation (23), § 39, Chapter II, an argument similar to that employed in § 1, we see that if ϕ be the angle turned through by one end of the wire under the action of a couple G, and if n be the rigidity,

$$\phi = \frac{2GL/m}{\pi n a_1^4} + \frac{2GL/m}{\pi n a_2^4} + \ldots .$$

The method of § 1 then leads to the equation

$$n = \frac{2GL}{\pi a_0^4 \phi} \left\{ 1 + \frac{10\Sigma b^2}{m a_0^2} \right\}. \quad \ldots\ldots\ldots\ldots\ldots\ldots\ldots(2)$$

The student who desires to obtain an intimate knowledge of all the circumstances of the experimental work may profitably determine from his observations the values of the correcting factors in (1) and (2).

NOTE VII.

On Inertia Bars.

The simplest way of attaching an inertia bar to a wire is to solder the wire into a small hole drilled in the bar, but it is generally more convenient to employ some method which allows the wire to be easily detached from the bar. A good plan is to solder each end of the wire into a hole drilled along the axis of a metal cylinder 2 or 3 cm. in length and 0·4 or 0·5 cm. in diameter. One of these cylinders fits easily into a hole drilled at the centre of the inertia bar at right angles to its length, and the cylinder is secured there by a small set screw, while the other cylinder is secured in the same manner in a hole drilled in a piece of metal held in a fixed support. The arrangement is illustrated in Fig. 33, Chapter III, § 62.

Since both ends of the wire are soldered into cylinders, the length of wire under torsion is quite definite, and since the torsional stiffness of the cylinders is very great compared with that of the wire, the couple due to a given angular displacement of the bar is practically independent of the positions of the cylinders in the bar and in the support, and thus no exact adjustment of the cylinders in the two holes is necessary.

The mass of the inertia bar should be determined *before* the hole is bored in it and, for convenience, the mass should be stamped or engraved on the bar. For a bar not less than 30 cm. in length the moments of inertia of the bar before and after the hole has been drilled in it do not differ appreciably from each other since the distance from the axis of the hole of every part of the metal which initially filled the hole was very small, while large parts of the bar are at considerable distances from that axis.

In the case of a rod of square section, 40 cm. in length and 1 cm. in breadth and depth, formed of metal of density 8 grammes per c.c., the mass of the bar is $8 \times 1 \times 1 \times 40$ or 320 grammes. If the moment of inertia about an axis through the centre at right angles to one of the larger faces be K_0 grm. cm.^{2}, we have by § 6, Note IV,

$$K_0 = \tfrac{1}{3} \times 320 \{20^2 + (\tfrac{1}{2})^2\} = 42693 \cdot 33 ... \text{grm. cm.}^2$$

Suppose, now, a hole 0·4 cm. in diameter is drilled in the bar, the axis of the hole coinciding with the axis just mentioned. The mass of metal removed is $8 \times \pi \times 1 \times 0 \cdot 2^2 = 1 \cdot 005$ grms. and by § 11, Note IV, the moment of inertia of the metal removed is

$$k = \tfrac{1}{2} \times 1 \cdot 005 \times 0 \cdot 2^2 = 0 \cdot 0201 \text{ grm. cm.}^2$$

If the moment of inertia of the bar after the hole has been drilled be K, then $K = K_0 - k$, and this, it will be seen, differs from K_0 by less than one part in two millions.

Though boring the hole has not appreciably affected the moment of inertia of the bar, it has changed the mass of the bar from 320 grms. to 320 − 1·005 grms. i.e. by one part in 320.   If we had taken 320 − 1·005 grms. as the mass of the bar and had treated the bar as a uniform rectangular block, the moment of inertia would be

$$\tfrac{1}{3} \times \{320 - 1\text{·}005\} \{20^2 + (\tfrac{1}{2})^2\},$$

and this would be less than K by one part in 320.

Sometimes a small stud is screwed into the inertia bar and the wire is secured by a set screw in a hole drilled in this stud.   It will be seen, from what has been said above, that in this case, also, the mass to be employed in the calculation of the moment of inertia is the mass of the bar before any holes are drilled in it and before the stud is attached to it.

NOTE VIII.

Work done by a Couple.

When a couple G dyne-cm. acts upon a body and the body turns through an infinitesimal angle $d\theta$ radians, an amount of work dW ergs will be done by the couple. We require to know how dW depends upon G and upon $d\theta$. We may suppose that the couple is applied by means of two strings A, B (Fig. 61) wrapped round a wheel of radius r cm. If the tension of each string be F dynes, we have

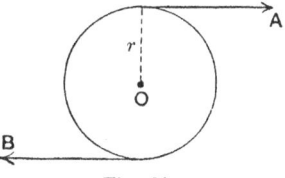

Fig. 61.

$$2Fr = G. \dots\dots\dots\dots\dots(1)$$

If, now, the wheel turn through $d\theta$ radians, the points A, B will move through $r d\theta$ cm. and each force will do $F \times r d\theta$ ergs.   The total work done is $2Fr d\theta$, and this is equal to dW.   Hence, by (1), we have

$$dW = G\, d\theta \text{ ergs.} \dots\dots\dots\dots\dots\dots\dots\dots\dots\dots\dots(2)$$

Thus the work done by a couple in turning a body through an infinitesimal angle is the product of the couple and the angle.

When the couple is constant, we have

$$W = G\theta \text{ ergs,}$$

where W is the work done by the couple while the body turns through the angle θ radians.

When the couple is proportional to the angle already turned through by the body from an initial position, we may write $G = \mu\theta$, and then the work done while θ increases from zero to ϕ radians will be the product of ϕ and the average value of the couple. The latter is $\frac{1}{2}\mu\phi$, and thus the work done is

$$W = \phi \times \tfrac{1}{2}\mu\phi = \tfrac{1}{2}G_{max} \times \phi \ \text{ergs},$$

where G_{max} dyne-cm. is the maximum value of G, i.e. the value of the couple when $\theta = \phi$.

NOTE IX.

Trapezoidal Rule for the Measurement of Areas.

When an area is enclosed by a line of some simple geometrical form, such as a triangle or an ellipse, the area can be calculated with any required degree of exactness by the integral calculus or other mathematical methods when the necessary dimensions are accurately known. But in practical work it is often necessary to determine approximately the area enclosed by a line drawn on paper, of which the whole or a part passes as evenly as possible among the points representing a number of observations. In such a case much labour would be involved in the attempt to determine, even approximately, the equation to the line, and then the calculation of the area by aid of the equation would still remain to be made. This method is, therefore, seldom used.

The area can be measured mechanically by means of a planimeter, but the accuracy of the result depends upon the correct adjustment of the instrument and upon the skill with which the tracing point is made to move along the line.

The trapezoidal rule for the measurement of areas is easily applied and requires no special instrument. In one respect it has an advantage over the planimeter method, for, when the observations are properly spaced, it is not necessary to draw the curve on paper.

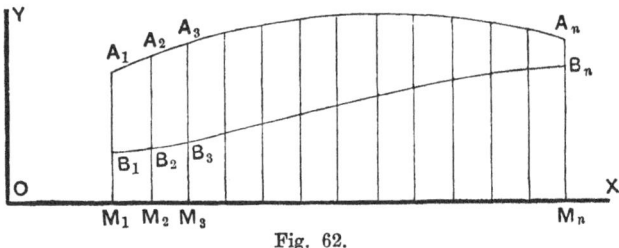

Fig. 62.

When we wish to find the area enclosed by the curve $A_1 \ldots A_n$ (Fig. 62), the axis of x and the two ordinates A_1M_1, A_nM_n, we divide M_1M_n into $n-1$ equal parts, each of length d cm. Let $A_1M_1 = a_1$ cm., $A_2M_2 = a_2$ cm. and so on. If d be small compared with the least radius of curvature at any point on the curve $A_1 A_n$, we may replace the arcs $A_1 A_2$, $A_2 A_3$ ... by the corresponding chords, and treat each of the vertical strips as a trapezoid. The area of the trapezoid $A_1 M_1 M_2 A_2$ is the product of $M_1 M_2$ and of $\frac{1}{2}(A_1 M_1 + A_2 M_2)$, the mean height of the chord $A_1 A_2$, and thus the area is $(\frac{1}{2}a_1 + \frac{1}{2}a_2) d$ square cm. The area of the next trapezoid is $(\frac{1}{2}a_2 + \frac{1}{2}a_3) d$ and so on. On addition, we find that, if the area $A_1 M_1 M_n A_n$ be A square cm.,

$$A = (\tfrac{1}{2}a_1 + a_2 + \ldots + a_{n-1} + \tfrac{1}{2}a_n) d. \quad\ldots\ldots\ldots\ldots\ldots\ldots(1)$$

The rule implied in (1) may be expressed as follows :—

Draw a series of equally spaced ordinates, add half the first and half the last ordinates to the sum of the intermediate ordinates and multiply the whole by the distance between successive ordinates. The result is the area required.

If we have a second area enclosed by the curve $B_1 \ldots B_n$, the ordinates $B_1 M_1$ and $B_n M_n$ and the base $M_1 M_n$, and if the $n-1$ equidistant ordinates be $b_1, b_2 \ldots b_n$ cm., the enclosed area B is given by

$$B = (\tfrac{1}{2}b_1 + b_2 + \ldots + b_{n-1} + \tfrac{1}{2}b_n) d. \quad\ldots\ldots\ldots\ldots\ldots(2)$$

If we denote $a_1 - b_1$ by c_1 and so on, and if C be the area $A_1 B_1 B_n A_n$, we have, by (1) and (2),

$$C = (\tfrac{1}{2}c_1 + c_2 + \ldots + c_{n-1} + \tfrac{1}{2}c_n) d. \quad\ldots\ldots\ldots\ldots\ldots\ldots(3)$$

In many instances (*e.g.* Fig. 55) the point B_1 coincides with A_1 and B_n coincides with A_n. Then $c_1 = 0$ and $c_n = 0$ and the formula (3) becomes

$$C = (c_2 + c_3 + \ldots + c_{n-1}) d. \quad\ldots\ldots\ldots\ldots\ldots\ldots(4)$$

By taking d small enough, the accuracy can be made as great as may be desired, provided that the values of the a's, the b's or the c's are *exactly* known.

If the observations be taken at equal intervals with respect to the variable quantity represented along the axis OX, it is unnecessary to *draw* the curve on paper, for it is only the values of the a's, b's or c's which we require and these are given by the observations.

When the observations are not taken at equal intervals with respect to the quantity represented along OX, we can either find the sum of the areas of the separate trapezoids corresponding to successive intervals (if the intervals be not too great), or we may draw the curve as evenly as we can among the plotted points and then find its area by the trapezoidal rule.

NOTE X.

HINTS ON PRACTICAL WORK IN PHYSICS.

1. FAILURES. A demonstrator in practical physics spends a large part of his time in correcting students' mistakes. He has to discover, for instance, why it is that a student obtains 537·86402 [no units mentioned] for Young's modulus by an experiment on a brass wire instead of $9·86 \times 10^{11}$ dynes per square centimetre. It is then found, perhaps, that the student has confused the radius of the wire with its diameter, that, having got hold of a screw-gauge in which one turn is equivalent to $\frac{1}{10}$ inch, he has treated one turn as equivalent to $\frac{1}{2}$ millimetre either because it looked about $\frac{1}{2}$ millimetre when tested with a millimetre scale or because he did not care to ask those who knew, that he has measured the extension in millimetres and has then treated the millimetres as if they were centimetres and that he has used 32 for "gravity" instead of 981. When the crumpled sheet of paper has been unearthed from the rubbish box, the arithmetic on it is found to be faulty. The student has omitted (perhaps through caution) all reference to the units in which the result is expressed. In some cases the student adds the letters C.G.S. in much the same way as grocers add "ESQ." to customers' names. If his courage allows him to name the units, he often uses the wrong names ; the chances are that he puts down "dynes."

The student may have learned something of the physical principles involved in the experiment and may have gained some practice in manipulation, but the result of his work, viz. that Young's modulus for brass is 537·86402, is worthless, and is entirely useless to any human being.

The following hints may perhaps assist the student to avoid errors in his work and may help him to discover where they have occurred when, in spite of all his care, his result is obviously wrong.

2. OBSERVATIONS. After the necessary adjustments have been made, the observer reads off a number from the graduations of the instrument or in other ways. The result of the experiment cannot possibly be correct if this number be not correctly read and correctly recorded. After the reading has been entered, the student should, when possible, look at the instrument again in order to detect any discrepancy between the *written entry* and the instrumental reading. What he *actually wrote* is not always what he *intended* to write.

The work of observing is liable to a great variety of errors. Some of the most frequent are the following :—

Wrong values are assigned to the divisions of a scale. Thus the student sees a 10 and counts on 5 more divisions, and enters the reading as 10·5 instead of 15. Or, when the main divisions are subdivided into 5 subdivisions, one of the latter is taken as a tenth instead of a fifth of a main division.

The numbered divisions are read from left to right, but the tenths are read from right to left. Thus 25·4 is wrongly read as 25·6, the 6 tenths in the latter number being reckoned from the " 26."

The student does not understand the graduation of the instrument, either because he has not given sufficient attention to the matter or because the unit of measurement is not marked on the instrument; in the latter case he cannot be expected to know the unit of measurement and he should ascertain it from those who have put the instrument into his hands, be they instrument makers, teachers, or examiners.

In most cases the determination of a physical quantity involves *two* observations. Thus, when the diameter of a wire is measured by a screw-gauge, the reading of the gauge when the jaws are in contact is required as well as the reading when the gauge is adjusted to the wire. But students frequently omit to take the zero reading. They should remember that " every length has two ends." The attempt to measure a length by a *single* reading sometimes leads to totally erroneous results, as when a distance of 30 cm. is put down as 70 cm. because the " wrong end " of the scale is used and so the distance to be measured lies between the " 100 " and the " 70 " on the scale, and not between the " 0 " and the " 30." If, in addition to the reading " 70," the reading " 100 " had been taken and *recorded*, the error would not have occurred. Similar remarks apply to the measurement of many other quantities, *e.g.* masses, angles, and resistances.

In finding the periodic time of a vibrating system, a student sometimes calls " one " when he starts the stop-watch; he stops the watch as he calls " fifty " and though he imagines that he has found the time of 50 vibrations, he has really found the time of only 49. He should call " nought " when he starts the watch.

When the periodic time exceeds about two seconds, the mind has time to ramble off to other interests between one count and the next, and therefore a special effort must be made to concentrate the attention on the work in hand. It is of assistance to count *out loud*. On account of the difficulty of counting correctly, the student should make at least *two* independent observations of any periodic time.

A steady hand, a keen eye, and a good general command of the body are essential in accurate physical determinations ; mere intellectual power avails nothing by itself. Any rule of life which deviates from temperance in all things (including work) may be expected to render the hand less steady and the eye less keen, and so to lead to inferior work. University students whose fingers are deeply stained with tobacco do not, as a rule, become skilful observers, though they may show considerable ability in other ways.

3. THE RECORDING OF OBSERVATIONS. As soon as an observation has been made, enter the result in a note book, not on a scrap of paper. Do not

wait to see the result of a second adjustment before recording the result of the first one. Take the figures *as they come* without any attempt to force them into agreement with any preconceived value.

Enter *observations* and not merely *deductions*. Thus, if two readings of a vernier be 15·85 cm. and 17·32 cm., these are observations. The distance 1·47 cm., through which the vernier has been moved, is a deduction from the two observations. If the student, without entering the numbers 15·85 and 17·32, does the arithmetic in his head and puts down 1·57 through error, he has no chance of detecting the mistake afterwards. If he had entered 15·85 and 17·32, he might have found the mistake in revising his calculation.

The neglect of the simple rule of always *entering* observations before making any deductions from them is a very frequent source of error. No one, whatever his private opinion as to his own powers, is likely to do reliable work if he neglects this rule.

Enter the observations in an orderly manner without crowding, **and** do not write in three or four different directions on the paper.

Write all the numbers very plainly. The letters in a badly written word can often be guessed, but the neighbouring figures do not help the reader to decide whether the mark on the paper is meant to be a 5 or an 8. The position of the decimal point is the most important feature of any collection of figures; be careful, therefore, to mark the decimal point firmly and clearly.

Be careful to state clearly what it is that you have measured, and also the units in which the measurement is expressed.

If you have reason to reject any of your observations, cancel the entries by bold lines drawn through them, so that there may be no mistake as to what is rejected and what is retained. Neatness is here of secondary importance.

A beginner naturally believes that he is capable of making a correct copy of the results of a series of observations; he will learn by experience that, in spite of his most strenuous efforts, mistakes *will* occur. It is therefore essential that the student should cultivate the habit of making the original record of the observations good and clear, and that he should preserve it for reference. If any practical use is to be made of the results of an experiment, it is obviously important that the chances of error should be as small as possible. The power of entering observations in a clear manner will be of value in a practical examination, for the student will then be able to send in his original record and will not feel compelled to waste time by copying out his "rough" notes.

4.   ARITHMETICAL REDUCTION OF OBSERVATIONS. From the observations the result is deduced by arithmetical work. Without this work the result cannot be obtained, and the accuracy of the result depends upon that of the arithmetical work. This work should therefore be carried out with quite as much care as that given to the taking and recording of the observations. The arithmetic should be done in the book containing the observations, and

the work should be arranged in an orderly manner so that it will bear inspection. It is wise to verify each step before proceeding to the next. Many students have the bad habit of doing the arithmetic on scraps of paper which they immediately destroy, as if they were ashamed of the work; yet no one expects them to obtain the results without doing the arithmetic.

For most purposes four-figure mathematical tables may be used; Bottomley's tables are convenient. The student should make himself acquainted with the contents of the book of tables so that he may know where to look for (say) the reciprocal of a number; and he will then not waste time in working it out by the aid of logarithms.

The slide rule is so convenient in those cases where moderate accuracy suffices, that the student should endeavour to become proficient in its use. But it must be recognised that its accuracy is limited.

Care should be taken to carry the arithmetic to a *sufficient* number of significant figures. The final result depends, of course, upon the data used in the calculations, but the arithmetic should be carried so far that no error is introduced into the result greater than (say) one tenth of that arising from the errors of observation. An example will make this clear. The value of the product $1·6736 \times 2·7628$ is $4·62382208$, or to 5 significant figures $4·6238$. But if we perform the multiplications, we find that

$$1·7 \quad \times 2·8 \quad =4·8 \quad \text{to 2 figures}$$
$$1·67 \quad \times 2·76 = 4·61 \quad \text{to 3 figures}$$
$$1·674 \times 2·763 = 4·625 \text{ to 4 figures.}$$

Hence the rough 2 figure arithmetic has introduced an error of about one in 25. With 3 figure arithmetic the error is reduced to about one in 330, and with 4 figures the error is only about one in 4000.

On the other hand, it is useless to retain many significant figures in the arithmetic when the data are only correct to a few significant figures.

When the number of significant figures is to be reduced by rejecting the last digit L, the last but one is left unchanged when L is less than 5, and is increased by unity when L is greater than 5. When L is *equal* to 5, the last digit but one is left unchanged if it is even, but is increased by unity if it is odd. Thus $3·485$ is shortened to $3·48$, but $6·235$ is shortened to $6·24$; in each case the number adopted after the rejection of the "5" has its last digit *even*.

When the numbers are very great or very small, it is best to write them thus :—$4·19 \times 10^7$ or $5·89 \times 10^{-5}$, keeping *one* significant figure only on the left of the decimal point. There is less chance of error in copying $5·89 \times 10^{-5}$ than in copying $0·0000589$. This plan has the advantage that, when the logarithms of the numbers are to be found, there is no need to count the number of figures between the decimal point and the first significant figure. The power to which the 10 is raised is equal to the characteristic of the logarithm. Thus

$$\log (4·19 \times 10^7) = 7·6222, \quad \log (5·89 \times 10^{-5}) = \bar{5}·7701.$$

The value of π is $3\cdot14159265\ldots$. It is quicker to use $\log 3\cdot141\ldots$ than to use $\log 22$ and $\log 7$, as is necessary when the rough value 22/7 is employed.

Gross errors in arithmetic can often be detected by the exercise of a little common sense. Thus a moment's thought shows that the cross-section of a wire one-tenth of a centimetre in diameter is *not* $2\cdot345$ square centimetres. The student should make a practice of looking at the result of each step to see if it is reasonable or absurd.

5. DIAGRAMS. When series of observations are plotted on squared paper, the student should express very clearly upon the diagram the two physical quantities which are represented along the horizontal and vertical axes. When this information is not given, the diagram is generally worthless. The points plotted on the diagram should be clearly marked by small circles drawn round them or in other ways.

In every case when a series of observations is made, one quantity X is varied and the consequent variations of a second quantity Y are observed. The quantity X should be varied over the whole of the available range and the separate values of X—say X_1, X_2 ...—should be fairly distributed over that range. Many students are inclined to take X_1, X_2 ... so close together that they are unable, for lack of time, to cover more than a small part of the whole range. In such cases, it often happens that the errors of observation cause the points plotted on the X-Y diagram to be suggestive rather of a constellation than of any regular curve. If the intervals $X_2 - X_1$, $X_3 - X_2$, etc., had been large, the errors of observation would not have completely obscured the law which the experiment was designed to investigate.

6. **NOTE BOOKS.** The student should, if possible, keep a note book in which to write fuller accounts of the experiments than is possible in the laboratory. He will thus find out how much he has understood of what he has done in the laboratory, and will also gain practice in describing experimental work in his own words. The note book should have large pages, and ample space should be left for future notes and additions. But however great the labour spent upon this book, it can never take the place of the laboratory note book in which the *original* records are written.

The student should write his name and address in his note books as a safeguard against their loss.

7. **GENERAL REMARKS.** The student should not leave an experiment while there is anything connected with it which he does not understand. Every experiment involves many principles, and thus a single experiment thoroughly grasped in all its details puts the student in possession of much knowledge which will help him in future experiments. Hence, one experiment well understood is of far more educational value than a dozen in which the student has gained only hazy notions.

There is no such thing as the ANSWER to any experimental investigation, for no two persons would obtain *precisely* the same result, however carefully they worked. The student should have confidence in his results until he discovers an error in his work. But he should not pretend to do impossibilities. It is easy to make some measurement, such as weighing, with a great show of precision, but the precision is only apparent and not real unless the proper precautions have been taken and the proper corrections have been applied.

As the degree of exactness to be reached in any measurement is increased, the practical difficulties increase enormously. Thus with a household balance and household weights a cook could weigh a mass of aluminium of about 100 grammes to one gramme. A junior student with a cheap laboratory balance and common weights could weigh it to $\frac{1}{10}$ gramme. To be certain of the mass to $\frac{1}{100}$ gramme, it would be necessary to use double weighing and to allow for the buoyancy of the air. To reach an accuracy of $\frac{1}{1000}$ gramme, it would be necessary to have a table of corrections for the weights employed, while to come within $\frac{1}{100.000}$ gramme would require an accurate knowledge of the pressure, the temperature and the hygrometric state of the air, and would require the refined appliances of a national physical laboratory and the skill of an expert.

The student should have an eye to *proportion*. It is useless to make some observations (*e.g.* of mass) to one part in ten thousand when other observations in the same experiment can only be made to one part in a hundred (*e.g.* rise of temperature).

The formula which expresses the result in terms of the quantities to be observed should be carefully examined to see which quantities are of primary and which are of secondary importance. Thus the formula

$$K = M \left(\tfrac{1}{3} L^2 + \tfrac{1}{4} A^2 \right)$$

for the moment of inertia of a cylindrical rod of mass M, of length $2L$ and of radius A, shows that, when L is great compared with A, the quantities M and L are of primary importance, while A is of only secondary importance. It is useless to spend time in measuring $2A$ accurately by means of a screw-gauge when $2L$ is only measured to the nearest millimetre, for it is, at the outside, only the first two significant figures in $\tfrac{1}{4} A^2$ which are of any consequence compared with $\tfrac{1}{3} L^2$.

NOTE XI.

MAXIMUM ELEVATION AT CENTRE OF ROD IN EXPERIMENT 6.

If $2L$ be the whole distance between the points HK (Fig. 35), $p=L-l$. Then, by (6), § 66,

$$h=(Mg/2EI)\,(L-l)\,l^2.$$

Hence $dh/dl=(Mg/2EI)\,(2Ll-3l^2)$, and $dh/dl=0$ when $3l=2L$. When $3l=2L$, $d^2h/dl^2=-MgL/EI$, and d^2h/dl^2 is negative. Hence h is a maximum when $l=\frac{2}{3}L$, or when $p=\frac{1}{3}L$. It is therefore advantageous to make l approximately equal to $\frac{2}{3}L$.

NOTE XII.

THEORY OF INFINITESIMAL UNIFORM BENDING OF A ROD.

If a rod of uniform section be uniformly bent, each filament which was parallel to the length of the rod when the rod was straight is bent into an arc of a circle. All these circles are in parallel planes—planes of bending—and their centres of curvature lie on a single straight line normal to these planes; this line is called the axis of bending. The uniformity of bending also requires that all the particles, which lay in transverse planes before the rod was bent, lie after the bending in planes through the axis of bending. These planes therefore cut the curved filaments at right angles. Hence the stress on any transverse section of any longitudinal filament is normal to that section and is thus a positive or negative tension.

Let the tension of a longitudinal filament be T dyne cm.$^{-2}$ and let ρ' be the radius of curvature and α the cross-section of the filament. Then the forces $T\alpha$ at the ends of an element of length $\rho'\theta$ of the filament are inclined at the elementary angle θ to each other and so give rise to a radial force $T\alpha\theta$, which is *towards* the centre of curvature when T is a positive tension. Hence the radial force per unit length of the filament is $T\alpha\theta/\rho'\theta$ or $T\alpha/\rho'$, and the radial force per unit volume is T/ρ'.

To maintain the filament in equilibrium, it is necessary that a radial force T/ρ' per unit volume should act upon the filament outwards from the centre of curvature. Except for any small effect of gravity, this force is supplied by the action on the filament of the contiguous filaments. We cannot treat the sides of the filament as free from stress unless this radial force is either zero or negligible. The importance of the vanishing of the radial force lies in the fact that only when the sides of the filament are free from stress can we write

$T = Ee$, where E is Young's modulus and e is the elongation of the filament, *i.e.* its increase of length per unit length.

When the rod is uniformly bent, the forces acting across any transverse section must be equivalent to a couple which has the same axis and the same magnitude for all such sections, and thus the resultant pull across any transverse section vanishes. In this case T and the curvature $1/\rho'$ diminish together. Hence T/ρ' diminishes on two counts as $1/\rho'$ diminishes. We may therefore assume that, as the bending tends to zero, the effects of the stress on the sides of the filament due to contiguous filaments tend to vanish in comparison with the effect of the longitudinal tension. We shall investigate the bending of the rod on this assumption.

To keep the matter simple, we suppose that the plane of bending through the "centre of gravity" of a transverse section cuts that section in a line which is an axis of symmetry for the section.

Let Fig. 63 be a projection of lines upon a plane of bending and let the axis

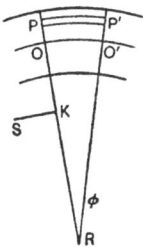

Fig. 63.

of bending meet the plane in R. Those filaments which are farthest from R are lengthened and those nearest to R are shortened. Hence some intermediate filaments are unchanged in length, and all these have the same projection. Let OO' be the projection of a filament of unchanged length and let RO, the radius of curvature of OO', be ρ. Let PP' be the projection of any other filament, let $RP = RP' = \rho + y$, and let POR, $P'O'R$ be radii. If $ORO' = \phi$, we have $OO' = \rho\phi$ and $PP' = (\rho + y)\phi$. Hence, if the elongation of PP' be e, we have

$$e = \frac{PP' - OO'}{OO'} = \frac{(\rho + y) - \rho}{\rho} = \frac{y}{\rho}.$$

The sides of the filament are, it is assumed, free from stress and hence, if the tension of the filament be T dyne cm.$^{-2}$,

$$T = Ee = Ey/\rho.$$

If the resultant of the pulls of all the filaments across the transverse section be F dynes,

$$F = \Sigma Ta = (E/\rho) \, \Sigma ay.$$

Since the forces applied to the rod on either side of the transverse section are equivalent to a couple and so have no resultant, the force F vanishes. Hence

$\Sigma ay = 0$, and therefore O is the projection of the "centre of gravity" of that transverse section which cuts the plane ORO' in OR.

The "bending moment" corresponding to the transverse section through POR is the moment about any axis parallel to the axis of bending of the forces which the filaments on the right side of POR exert upon the part of the rod to the left of POR. For convenience we take the axis through O. The pull along PP' is Ta or Eay/ρ and, since $PO = y$, its moment about the axis is Tay or $(E/\rho) ay^2$. If the total moment be G dyne cm.,

$$G = (E/\rho)\, \Sigma ay^2 = EI/\rho,$$

or

$$G\rho = EI,$$

where $I = \Sigma ay^2$. Here I is the "moment of inertia" or the "second moment" of the area of a transverse section about an axis through the "centre of gravity" of the section parallel to the axis of bending.

Instead of taking the axis through O, we may take it through any point S. If SK be perpendicular to PR and if $KO = k$, the moment is

$$(E/\rho)\, \Sigma\, \{ay\,(y+k)\}.$$

But $\Sigma ay = 0$, and thus the moment reduces to EI/ρ and is independent of the position of S.

The tension T in a longitudinal filament gives rise to a force T/ρ' per unit volume, and the reaction to this force necessary for equilibrium is supplied by the contiguous filaments. Since $T = Ey/\rho$ and $\rho' = \rho + y$, the force is

$$Ey/(\rho^2 + \rho y),$$

which tends to become inversely proportional to ρ^2 as ρ increases. The tension is inversely proportional to ρ. The effects of T/ρ', the force per unit volume, can be made as small as we please in comparison with those of the tension T by making the bending sufficiently small. We conclude that $G\rho$ tends to the limit EI as the curvature $1/\rho$ tends to zero.

We have not proved and it is not, in general, true that $G\rho$ *equals* EI when the curvature is *finite*. The problem of small *finite* bending is considered in §§ 27—38.

INDEX.

The references are to pages and not to sections.